JOURNEY TO A REVOLUTION

BY MICHAEL KORDA

Ulysses S. Grant

Marking Time

Horse People

Making the List

Country Matters

Another Life

Man to Man

The Immortals

Curtain

The Fortune

Queenie

Worldly Goods

Charmed Lives

Success

Power

Male Chauvinism

BY MARGARET AND MICHAEL KORDA

Horse Housekeeping

Cat People

HarperCollins*Publishers*

Journey to a Revolution

A Personal Memoir and History of the Hungarian Revolution of 1956

MICHAEL KORDA

HarperCollins books may be purchased for educational, business, or sales promotional use. For information, please write: Special Markets Department, HarperCollins Publishers, 10 East 53rd Street, New York, NY 10022.

FIRST EDITION

Designed by Amy Hill

Frontispiece: author (far left), Roger Cooper, Russell Taylor, and Christopher Lord in Budapest, October 1956. Credit: Erich Lessing / Magnum Photos

LIBRARY OF CONGRESS CATALOGING-IN-PUBLICATION DATA

Korda, Michael, 1933–
Journey to a Revolution : a personal memoir and history of the Hungarian Revolution of 1956 / by Michael Korda.
p. cm.
Includes bibliographical references and index.
ISBN-10: 0-06-077261-1
ISBN-13: 978-0-06-077261-1
1. Hungary—History—Revolution, 1956. 2. Europe, Eastern—History—1945–1989. 3. Communism—Europe, Eastern. I. Title.
DB957.K674 2006
943.905'2—dc22 2005058170

06 07 08 09 10 ❖/RRD 10 9 8 7 6 5 4 3 2 1

For Margaret, with love—

and for Morton L. Janklow,

my friend of more than four decades,

for his enthusiasm and our shared

love of history

Talpra magyar,
hí a haza!
itt as idő,
most vagy soha!

—Petőfi Sándor, *Nemzeti dal*

Arise, Magyars,
The country calls!
It's now or never,
What fate befalls . . .

—Sándor Petőfi, "National Poem"

CONTENTS

PREFACE

Nothing presents more difficulties than writing an objective account of a great event in which one has participated oneself. Being objective about the Hungarian Revolution of 1956 is hard enough to begin with—it is a kind of modern David and Goliath story, except that Goliath won—but harder still for somebody who participated in it, like myself, and who was present at some of the major moments. The *abbé* Sieyès, when asked what he did during the French Revolution, replied, *Je l'ai survécu* ("I survived it"). I suppose I can, at the very least, make the same claim, but I also witnessed demonstrations of almost unbelievable courage, of shameless political betrayal, and of the kind of unrestrained military brute force and violence against civilians of which the twentieth century was so full, and which do not seem to have ended so far in the early years of the twenty-first.

Although there is no doubt that the events of October 1956 constituted a revolution, they swiftly became something else: a war.

The revolution was spontaneous, popular, and embraced every segment of society, including many members of the Hungarian communist party. It was a revolution against tyranny—against a singularly brutal and despotic single-party government that punished even the mildest dissent with extreme cruelty, including torture and death; that methodically stifled every form of self-expression and suppressed every individual impulse; and that was more Stalinist than Stalin himself had been.

It was also a revolution against eleven years of alien, heavy-handed, unyielding Russian domination and occupation. The Hungarian Revolution of 1956 swept away the communist party's control over every aspect of Hungarian life, and for a time forced the army of the Soviet Union to retreat before an angry armed body of "Freedom Fighters," as they came to be known, consisting of workers, students, Hungarian army personnel, and ordinary civilians—men, women, and children. At that point the Soviet Union launched a full-scale war against the Hungarian people, suppressed the revolution, and handed the country back to the hard-core Hungarian communist bureaucrats and their feared and hated "security service."

Fifty years after the revolution, we are living in a very different world, one in which Hungary is a prosperous member of the European Union and of NATO; in which the Soviet Union has ceased to exist; and in which communism is virtually extinct except in China and Cuba. It is therefore important to remind ourselves of these events—for without the bravery of the men and women who fought in 1956, we might today still be living in the world of the cold war, of monolithic communism in eastern Europe, and of nuclear stalemate.

I hesitated for some time before deciding to include myself in this account of the revolution, but it seemed to me that many of these events benefit from being described by an eyewitness. I have tried to strike a balance between history and memoir, and have not tried to write a "first-person account," since there were of course many things I did not see or participate in, and since my own experiences in the streets of Budapest in 1956 were primarily those of an observer. Many others took far greater risks than I did, and it is in their memory that I have written this book.

Throughout, I have tried to set down as accurately as I can what took place and why without sensationalism. In the days while I was in Budapest many things happened of a personal nature which I have *not* described here—this book is about the Hungarians and their revolution, not about me.

"I am a camera," Christopher Isherwood wrote of his presence in Berlin in the years just before the Nazis came to power, "with its shutter open, quite passive, recording, not thinking. . . . Some day, all this will have to be developed, carefully printed, fixed."

So it was, and is, for me. I was a camera. I have recorded certain scenes as they were, fixed in my memory, like the black-and-white images of a war photographer, to give the reader an impression of what it was like to be in the middle of a revolution, and to show just what it is like to be on the spot when history and politics take to the streets.

But before one can explain the revolution, one must first explain what the cold war was like, in the days when the Iron Curtain was real, and still freshly built, dividing Europe in two as if it had been cut by a knife.

JOURNEY TO A REVOLUTION

1.

The Idol with Feet of Clay

Few things stand out less clearly at the time than a turning point in history, at any rate when one is living through it. As a rule it is only in retrospect that an event can be seen clearly as a turning point. Historians write as if they were looking at the past in the rearview mirror of a moving car; and, of course, picking the "turning points" of history is something of a specialty for many historians—in some cases, the more obscure, the better. Turning points, however, are much harder to recognize as they occur, when one is looking ahead through the windshield.

To take an example, we now recognize that the Battle of Britain was a turning point in World War II—fewer than 2,000 young fighter pilots of the Royal Air Force handed Hitler his first defeat, and ensured that whatever else was going to happen in 1940, Great Britain would not be invaded—but those who lived through the Battle of Britain day by day did not perceive it as a decisive, clear-cut event. The fighter pilots were too exhausted and battle-weary to care; and the public, while buoyed by the victories of the R.A.F. in the sky over southern England, was still struggling to come to grips with the evacuation of the British Expeditionary Force from Dunkirk and the collapse of France, and would soon be plunged into the first stages of what later came to be known as the blitz. Those who—as children—saw the white contrails of the aircraft swirling overhead in the blue summer sky, or watched the shiny brass cartridge cases come tumbling down by the thousands, had no sense of being witnesses to a "turning point"; nor did their elders. It was only much later that this turning point began to be perceived as one, and that Battle of Britain Day was added to the list of annual British patriotic celebrations.

In much the same way, the Hungarian Revolution of 1956, while it was clearly a major event, was not perceived as a turning point in history until much later, when the unexpected disintegration of the Soviet empire in eastern Europe, followed very shortly by the total collapse of the Soviet Union itself, could be traced back to the consequences of the uprising in the streets of Budapest.

The three weeks of the Hungarian Revolution ended, of course, in a victory for the Soviet Union, as everybody knows, but not since Pyrrhus himself has there been so Pyrrhic a victory. The Hungarians had chipped the first crack in the imposing facade of Stalinist communism and had exposed the Soviet Union's domination of eastern Europe for the brutal sham it was. For the first time, people in the West—even those on the left—had seen the true face of Soviet power, and it shocked them.

The Hungarians lost, but in the long run the Russians lost more. Communism became much harder to sell as a humane alternative to capitalism (or to western European democratic socialism); and the Russians themselves, badly shaken by the size and the ferocity of the uprising they had put down with such overwhelming force, and dismayed by the attention it received in the world's media, never attempted to repeat the experience in Europe. From time to time, the tanks might be sent rumbling into the streets again, as they were in Prague in 1968, but they would not henceforth open fire on civilians. Without apparently having given the matter much thought, the Russians discarded the trump card in their hand: the belief on the part of eastern Europeans that the Red Army would shoot them down mercilessly if they rose against the puppet governments the Soviet Union had imposed on them at the end of World War II.

More dangerous still, those governments themselves, whose ultimate legitimacy rested on the threat of armed intervention by the Soviet Union, ceased to believe that it would ever intervene again to support them with force the way it did in Hungary in 1956—and if the Russians would not, then how, when push came to shove, were the "people's governments" of Poland, Czechoslovakia, Hungary, the German Democratic Republic, Bulgaria, Romania, and Albania to remain in power over the long haul?

This is always a serious problem of empire, by no means limited to the Soviet Union. In 1919, the United Kingdom used violence, albeit on a considerably smaller scale, against the congress party in India when General Dyer ordered his troops to open fire on demonstrators in Amritsar, killing 379 of them at Jallianwallah Bagh, and wounding perhaps three times as many. The massacre horrified the British—except for the impenitent General Dyer and his supporters—and the result was an increasing reluctance on the part of the British government to use force against the congress party at all. In consequence, the threat of armed violence—one of the pillars on which the raj stood—gradually became more and more remote and unlikely. The Indians ceased to fear it, the

British grew increasingly unwilling to use it on a large scale, and independence for India thus became only a matter of time.

In much the same way, the Hungarians' uprising against their own unpopular government and the government's Soviet masters, while it failed, fatally shook the confidence of the Soviet government in its ability to control the countries of eastern Europe—a confidence that had already been weakened by Stalin's death, by the increasingly (and defiantly) independent attitude of Tito's Yugoslavia, and by the widening ideological rift between the Russian and the Chinese communist parties.

After 1956, the Soviet Union found itself in an increasingly uncomfortable and ambivalent position vis-à-vis the eastern European "people's democracies," since the Soviet leadership was desperately trying to dismantle the remnants of Stalinism at home and bring about a "thaw" in Russian life, while at the same time continuing to prop up unrepentantly Stalinist leaders in the Soviet Union's client states. In the Politburo there was a dawning realization that the Soviet Union was in a race against time to create a viable consumer economy while still maintaining and modernizing its huge armed forces, as well as to allow a greater degree of freedom of expression than had been thinkable under Stalin without altogether losing the party's ultimate control of the media.

This was a tricky balancing act for Stalin's contentious heirs, akin to a herd of elephants trying to walk a tightrope. While performing this miracle, the members of the Politburo did not want their attention to be diverted toward the "satellite countries," as the eastern European communist regimes were called in the West, or the "fraternal socialist countries," as they were called in the Soviet Union. It was the job of the eastern European leaders to maintain discipline and order in their countries—in short, to keep everything frozen, "on ice"—a job which was by no means easy to do when the Soviet Union itself was experimenting, however gingerly, with a thaw.

It was the unhappy lot of the Hungarians in 1956 to be the first people to seriously challenge Russian hegemony in eastern Europe—not just to challenge it intellectually, or with speeches, which was already dangerous enough, but to challenge it with armed force. The Russians' response was instinctive, brutal, and violent—in effect, to do exactly what Stalin would have done—but cruel and harsh as the suppression of the Hungarian rebellion was, it was not followed by the kind of widespread terror that would have come naturally to Stalin. Certainly, many Hungarians were shot, and many more imprisoned, but there were no mass repressions, no transfers of population to the gulags or to remote parts of the Soviet empire, no attempts to wipe out whole sections of the population.

Quite apart from the reluctance of the Soviet leadership to carry out such a program, two things had occurred that nobody had foreseen, both of which stemmed from the fact that Hungary has a common border with Austria and with what was then West Germany. One occurrence was that from the moment the Hungarian Revolution broke out it became possible to enter Hungary quite freely, with the result that this was the first event of its kind to be covered in detail from the beginning by reporters, television cameramen, and photographers from all over the world. Few historical events, in fact, have ever produced more copy and pictures than the Hungarian Revolution, and the pages of *Life, Paris-Match,* and *Stern* (and their counterparts all over the world) brought the revolution home to readers in searing and unforgettable images, none of them likely to create goodwill for the Soviet Union.

The second was that the open frontiers with the West made

it possible for large numbers of Hungarians to leave once it was clear that the revolution had failed—whole classes of people whom Stalin would have had shot or sent to the gulags simply walked across the frontier into Austria or West Germany, and from there moved on to the United Kingdom, Canada, and the United States. This Hungarian diaspora was not unlike the flight of educated, upper-middle-class Cubans to Miami after Castro seized power.

Hitherto, accounts of Soviet repression had been anecdotal, and many people on the left didn't believe them. The Chechens had been shipped in cattle cars in midwinter to exile in Siberia without their suffering having been recorded by photographers for *Life,* by cameramen for the network news shows, or by best-selling novelists; the gulags were out of sight (as well as, for most people in the West, out of mind); the murder of millions of kulaks during the collectivization process went undocumented; in short, the worst of Stalin's crimes had gone unseen and had left almost no record. But Budapest was full of eyewitnesses, and events there were on the nightly news and the front pages in the West, impossible to ignore. Not only that; all this was, from the point of view of the press, "great stuff": burning buildings; tanks blown up by heroic schoolchildren with Molotov cocktails; weeping, wounded babies in the hospitals; students fighting the Red Army in the streets. These events were perfectly calculated to make "news," like something out of the pages of Evelyn Waugh's *Scoop*. Here was everything Lord Copper could have wanted—heroism in the streets, the release of an imprisoned cardinal, the secret police shot down by angry civilians, and beautiful girls cradling submachine guns.

The hapless Russians were caught in the full glare of the Western news machine, blinking in the face of flashbulbs and growling with rage, like an adulterous husband caught in flagrante delicto by the Fleet Street press corps. As if that were not bad enough, the great numbers of Hungarians who escaped after the revolution—and who were speedily settled in the West, to assuage Westerners' guilt for not having intervened in or supported the revolution—kept the event alive in people's imagination for many years; so, unlike most news stories, this one did not fade quickly from people's minds. People had seen, with their own eyes, what Soviet repression was like; it was not quickly forgotten on either side of the Iron Curtain; and great sympathy was extended to those who had survived and escaped it.

When Talleyrand was told the news of Napoleon's execution of the duke d'Enghien and asked what he thought of what was then regarded by many people as a state crime, he replied, "It is worse than a crime—it is a blunder."

Much the same can be said of the Hungarian Revolution. It was a blunder of historic proportions on the part of the Soviet Union. The initial demands of the Hungarians could have been met, perhaps not totally, but partially—certainly enough to restore order. "Socialism with a human face," as it was later called, could have been put in place in Hungary and might have spread to the rest of the Soviet Union's eastern European empire, and the Warsaw Pact could have been turned into a genuine alliance. Instead, the governments of the people's democracies, for their own survival, were obliged to tighten the screws on their own people. The mailed fist was henceforth the only official political answer to any problem; and it became clear, even to those who had not hitherto been willing to believe it, that the only way forward to a better life would require the destruction of the whole communist system, and the complete removal of Russian armed forces from eastern Europe.

Jack Esten/Getty Images

The authentic face of Soviet communism: a Russian senior officer, unfastening his pistol holster, approaches Western newsmen.

An account of that event—undoubtedly a turning point in the full meaning of the phrase—seems appropriate fifty years after it occurred, when we can see, quite clearly, to borrow the words of Winston Churchill, that while it was not the end, or even the beginning of the end, it was, perhaps, the end of the beginning. The uprising in Budapest in October 1956 marked the first and most significant failure of the Soviet Union in its hard-won role as the dominant power in eastern Europe, as well as a monumental public failure in foreign relations (and public relations) for the heirs of Stalin. After 1956, Soviet policy in eastern Europe seesawed between timid attempts to soften the grasp of the party and the police, and to allow each country a little more freedom in terms of its own special historical institutions and values—for example, not trying too hard to eliminate private farming in Hungary, or turning a blind eye to the role of the Catholic church in Poland—and halfhearted threats of force, which achieved little except to keep alive the feeling among eastern Europeans that the Soviet Union, communism, and the Red Army were their enemies, and that their own governments were merely collaborationists and traitors.

This feeling was exacerbated by the fact that from the point of view of the Russians, the eastern Europeans, to paraphrase Harold Macmillan, "never had it so good." However grim, grimy, and impoverished life might seem to Western visitors in Prague, East Berlin, Budapest, or Warsaw, it was a good deal better than life in Moscow or Leningrad, let alone in the more remote industrial cities in the Soviet Union. The fact that agriculture had been only halfheartedly collectivized in the eastern European countries led to a greater abundance of food; besides, the eastern European countries had not experienced the full horrors of Leninism and Stalinism from 1917 on. Most of them still had a substantial educated middle class, however much it was shorn of political rights and property, and some tradition of bourgeois democracy. The party might try to discourage religious belief, but the churches had not been eradicated any more than the peasants had been turned into kolkhozniki (collective farmers). From the Russian point of view, the countries of eastern Europe had nothing to complain about. This made it all the more irritating that they did nothing but complain.

Worse, the increasing inability of the Soviet Union to hold its satellite states down and impose orthodox Stalinist-Leninist politics on them—the realization that nothing but brute force, brutally applied, could prevent the Hungarians and the Czechs from moving toward a free market economy; or the Poles from going to church and forming independent labor unions; or the East Germans from trying to escape to West Germany despite the Wall; or the Romanians from hating the ruling Ceaușescu family—eventually forced the Soviet Union to relax its hold on eastern Europe. Even a world power could do only so many things at once. The Russians were trapped in an extravagantly expensive arms race with the United States; sponsoring revolutions as far abroad as Central America and Africa; supporting a host of Middle Eastern "clients"; attempting to suppress rebellion in Georgia, Afghanistan, and most of their own southern border republics; and competing with the increasingly obdurate Chinese for leadership of the communist world. The Russians' economy was hopelessly overcentralized; their industrial base, except when it came to military production, was antiquated; agricultural production was declining; and the Russians themselves, in the dawning new age of satellite television and personal computers, were for the first time exposed to the temptations of the Western consumer economy—temptations which the planners in the Politburo were unable to stifle or to satisfy. The Soviet Union was thus in no position to use force in eastern Europe, even had it wanted to; and as the Soviet army bogged down in Afghanistan—the Russian equivalent of America's involvement in Vietnam—even confidence in the army began to wane. The Soviet Union was not yet a paper tiger, but it was in the process of becoming one, and its troops in the eastern European countries began to be pitied rather than feared.

It was Gorbachev's great and fatal miscalculation to assume that he could allow the eastern European countries to give up communism yet still preserve it at home. In fact, the collapse of communism in the Soviet Union's client states inevitably led to its swift disappearance in Russia as well. The system collapsed like a house of cards, with a swiftness that astonished everyone.

The Hungarians had been the first to grasp that the idol had feet of clay, and the first to try to destroy it—the toppling of Stalin's huge statue in Budapest in October 1956 was deeply symbolic—but the Russians, having suppressed the Hungarian Revolution, failed to learn a lesson from the experience. They expanded their role as a world power—a nuclear superpower, competing with the United States all over the world—without trying to solve the problems (or, in Marxist terms, the "contradictions") in their own backyard; and eventually ferment and dissent close to home fatally undercut the regime. It took twenty-five years for the Hungarians to win the battle they had begun (and lost) in October 1956, but in the end they did win. As Russians said to each other, with bitter irony, in 1988, while Soviet institutions collapsed around them, paraphrasing a favorite communist slogan, "For seventy years we have been on the road to nowhere."

. . .

This book will try to explain first of all how the Hungarian Revolution came about, and, of necessity, to explain to the reader how Hungary came to be in the position it was in as of 1956; and then to trace the events of 1956 both as history and in terms of my own experience there (which will likewise require a bit of explanation).

Like most significant historical events, the Hungarian Revolution did not take place by accident or in a vacuum; but neither was it carefully planned, as the Russians wanted the world to believe. In retrospect, of course, all things—or most of them—are clear, but the various strands that brought the Hungarians into armed conflict with the Soviet Union require a fairly patient unraveling if they are to be understood. Some things can probably never be completely sorted out—the importance of Radio Free Europe in encouraging the Hungarians to take up arms and to believe that the United States would support them, for example; or the involvement of the British government and MI6 in goading the Hungarians to fight in order to tie up the Soviet Union in the streets of Budapest while the United Kingdom, France, and Israel invaded Egypt—but I will do my best to distinguish fantasy from reality.

Countermyths have grown up too, of course, some of which were dreamed up by the KGB at the time of the uprising and have taken root and flourished over the decades—that the Freedom Fighters were right-wing anti-Semites, for instance; or that the entire episode was the work of Western intelligence agencies, a CIA plot from the start.

At the time, it looked on the contrary very much as if the Western powers were trying to calm the situation down and hold the Hungarians back, rather than supply them with weapons or help of any kind, but appearances in such matters can be deceiving. Certainly, the Hungarians were disappointed not to receive any concrete help from the West—particularly from the United States—and most better-educated Hungarians realized that their revolution was doomed the moment they heard the news of Suez, and were briefly embittered as a result against Great Britain and France; but on the whole, feelings toward the West still remained friendly even among Hungarians who felt they had been betrayed. In any case, as with every revolution, once it began it took on its own logic and moved at its own pace—there was simply no way to stop it or slow it down, even had anybody wanted to. It would end in victory or in defeat, and that was that.

To understand how and why the revolution began, of course, it is necessary to go farther back than 1956, or even 1945, when the Red Army finally took Budapest after a long and bitter siege and drove the German army out. Hungary lies on the border—one might almost say the fault line—between East and West, and always has. Settled by the descendants of Attila's Hun followers (hence "Hungary"), it eventually flourished as a Christian kingdom; was defeated and occupied by the Turks; and was liberated, only to become ruled from Vienna as a reluctant part of the Hapsburg empire.

Given its history and its geography, Hungary always looked toward the West and feared the East.

Hungary was a land of many brave, hopeless battles against overwhelming odds, so it was perhaps not surprising that the Hungarians took up battle against the Soviet Union at the height of the cold war.

2.

Hungary: The Mythic Nation and the Real One

In no country's history is it ever easy to distinguish myth from fact, but making that distinction is even harder than usual when it comes to Hungary.

To begin with, there have always been several superimposed Hungaries, each one quite different. First, of course, is the physical, geographical Hungary, a land of astonishing beauty and richness—though here too there is some confusion, for Hungary has grown and receded many times over the centuries, and thanks to the Treaty of Trianon in 1919 much of the land that Hungarians traditionally regard as theirs now lies in Romania, Slovakia, Croatia, and Poland, along with the Hungarians who still resentfully live there. It is hard for an American or a Briton to understand how difficult history is for a small country with no easily fixed or natural borders, with larger and much more powerful neighbors to the east and west, and with exalted territorial ambitions of its own against its smaller neighbors.

Second is the historical Hungary, a nation that is now 1,110 years old—as old, that is, as France. This Hungary was transformed from a way station of nomadic, mounted Hunnish tribes (inventors of the stirrup, beef jerky, and the wood-framed leather-covered saddle) moving west in the wake of Attila into a functioning monarchy under the leadership of Árpád. It struggled for its existence during most of its tumultuous history, eventually becoming the easternmost bastion of Catholicism until it was conquered by the Turks in 1526. Hungary's population fell from 4 million to 1.5 million under Turkish rule—the Turks were early practitioners of genocide—and Hungary was finally liberated in 1686 only to be engulfed by the Hapsburg empire, against which the Hungarians rebelled in the eighteenth and nineteenth centuries, with tragic consequences.

Third is the mythic Hungary, land of Gypsy music; dashing horsemen; beautiful, hot-blooded women; paprika; rich, spicy food; full-bodied wines; and Tokay.

Fourth is the Hungary of the diaspora—a nation with just over 10 million inhabitants, plus almost 3 million living—whether willingly or not—abroad.

The fourth Hungary has always commanded more attention than the others, for Hungary's greatest export has always been people. A good example: in the eighteenth century Moric Benyovszky, a Hungarian, fought with the Poles against Russia; was taken prisoner and deported to Kamchatka, in easternmost Siberia; managed to escape by seizing a Russian man-of-war and sailing it around the world, stopping among other places at Madagascar, which impressed him greatly; and then made his way to France, where he persuaded King Louis XV to let him establish a French colony in Madagascar. Louis XV, not an easy man to persuade, eventually made Benyovszky a count and allowed him to proceed, whereupon Benyovszky returned to

Madagascar, defeated the local tribes, and in 1776 had himself declared emperor of Madagascar—hardly what Louis XV had in mind. He persuaded those of the inhabitants of Madagascar who could read to adopt Latin script, with Hungarian spellings; made a visit to Britain and the United States seeking support; and was appointed a general by Empress Maria Theresa of Austria before losing his life in a coup staged against him by rival French colonists.

Benyovszky set the pattern for many Hungarian expatriates who followed him—one sees in him a fatal combination of courage, optimism, imagination, restless talent, charm (after all, he charmed both Louis XV and Maria Theresa, two supreme realists), and unmistakable *folie de grandeur* which would soon become only too familiar.

Benyovszky was not alone in his gift for charming the French king and court. Count Laszlo Bercsenyi, a gifted eighteenth-century cavalry officer, brought to France the idea of the *huszár*—the Hungarian light cavalryman, with a characteristic short, gold-laced, fur-trimmed, fur-lined cape thrown over one shoulder; a high fur cap; skintight embroidered riding breeches; and long, curved saber—thus changing the French cavalry, and eventually the cavalry of all the great European nations, which one after the other adopted the hussars as free-ranging light cavalry. Light cavalry had a completely different role and set of tactics from conventional heavy cavalry; the latter carried long straight swords and wore thick thigh-high boots, heavy helmets, and gleaming cuirasses. (It was the British light cavalry that charged so disastrously at Balaklava in the Crimean War, a role for which they were never intended. The heavy cavalry is still represented by the Household Cavalry, whom tourists see on mounted duty in London and at Windsor Palace.)

It must have taken a singular degree of charm—and what would later come to be known as chutzpah—to persuade the officers of what was then Europe's most powerful army to accept a major innovation in cavalry, always the most conservative of arms, from a small country far away on the Danube; it is pleasant to record that Bercsenyi came to a happier end than Benyovszky—he was created a marshal of France, the only Hungarian to reach the highest rank in the army of France.

History has always provided talented and ambitious Hungarians with good reasons to leave Hungary and seek their fortune abroad. In the eighteenth century Prince Rákóczi's abortive uprising against the Hapsburgs led many aristocratic Hungarians to flee. A century later the failure of the revolution of 1848–1849, which was crushed by the Austrians with help from the Russians, led many educated Hungarian professional men, as well as aristocrats, to leave.

There have been, in fact, five distinct periods of emigration: that of the eighteenth century; that which followed the failed rising of 1848–1849; a largely proletarian emigration to the United States toward the end of the nineteenth century, due to an economic crisis and disappointed expectations in the Hungarian working class; the emigration of the 1920s and the 1930s, mostly of well-educated, assimilated, successful Hungarian Jews who had the good sense to see that the fairly mild, institutionalized anti-Semitism of the Horthy government was likely to turn into something far worse, as indeed it did after Hitler came to power in 1933; and, finally, after the failed revolution of 1956, the mass escape of several hundred thousand Hungarians who left during the period when the borders were still open.

In each case a failed revolution or a political failure triggered a massive emigration (in terms of Hungary's relatively small population), but in each case it was a different class of people who left. The Hungarian aristocrats who crowded the courts of Europe in the eighteenth century were very different people from the middle-class professional men and army officers who emigrated after Kossuth's revolution (many of the latter reached high command in the Union army during the American Civil War). The working-class families that emigrated to America and settled in Ohio and Illinois to work in the coal mines and steel mills were very different from the middle-class, assimilated Jewish doctors, bankers, artists, film and theater people, journalists, and writers who fled from Horthy and later from the threat of Hitler's influence over Horthy. In 1956 the refugees included people of every social class who no longer wished to live under communism, or, in a smaller number of cases, had taken too visible a role in the revolution to risk remaining in Hungary at the mercy of the Hungarian or the Soviet secret police.

But Hungarian men of talent or genius had always emigrated, even when politics was not the cause. Although Budapest was a beautiful, sophisticated, cosmopolitan city, the capital of the Austro-Hungarian empire was Vienna. The court was in Vienna, as was the German-language press, and with it the possibility that one's fame might spread to the greater world beyond—to Paris or, later, Berlin. Hungary was small; its language was impenetrable. Eventually, with whatever regrets, it was difficult to resist taking the road that led to Vienna, and from there to the other capitals of Europe. The great Hungarian pianist and composer Franz Liszt took that path, and he was neither the first nor the last. Liszt found fame, but like many emigrants he always looked homeward with nostalgia. All Hungarians did. Easily assimilated as they were (Benyovszky spoke fluent German, French, Polish, English, and Russian, and was made a count by the Poles, by Louis XV, and by Maria Theresa—who also made him a general—before persuading the Madagascar tribes to declare him emperor), the Hungarians never lost their nostalgia for Hungary.

This nostalgic Hungary—a colorful folkloric land of wailing Gypsy violins; spicy goulash; frothy desserts; and great estates where exotically dressed aristocrats still hunted bison, wolves, and bear—became, thanks to generations of Hungarian emigrants, more or less fixed in people's minds. In much the same way, millions of emigrants from Ireland eventually succeeded in imprinting in people's minds a mythic Ireland of lush green fields, leprechauns, shamrocks, and smiling colleens that bore scant resemblance to the land from which they had fled to escape starvation, poverty, fratricidal politics, religious persecution, and 900 years of English occupation.

Perhaps fortunately, Hungary has no equivalent of the green plastic bowler hats and buttons that read "Kiss me, I'm Irish!" worn on St. Patrick's Day even by people who haven't a drop of Irish blood; but like Ireland, the mythic Hungary came to have a weightier significance in people's minds than the real one. People who couldn't name the capital of Hungary knew all about Transylvania, with its towering mountains and its packs of wolves wailing in the moonlight outside the sinister castle of Count Dracula—even though, thanks to the Treaty of Trianon, the vampire count, like so many more ordinary Hungarians, eventually found himself living in Romania.

The fact that Hungary in the twentieth century was a real country, with growing heavy industry, mines, and a rich base of agriculture, as well as numerous bitter grievances against its neighbors and insoluble internal political problems, passed almost unnoticed in the greater world. In part this was because there was a certain undeniable comic-opera quality to Hungary's situation: although Hungary is a landlocked country, it was ruled from 1919 almost to the end of World War II by an admiral. The national reputation for charm extended to producing a portrait of Hungary as a charming place, and playing down such problems as the fact that the Horthy regime was a dictatorship, or that it tolerated and even encouraged a high level of anti-Semitism. Even Jewish émigrés in the 1920s and 1930s continued to talk about Hungary with tears in their eyes, and seldom mentioned the reason they had fled abroad in the first place. Hungarians were adaptable; they had a natural ability for languages (hardly anybody but themselves in the Austro-Hungarian empire spoke Hungarian, so they were obliged to learn German if they wanted to get ahead); and they benefited from the one thing at which the Austro-Hungarian empire excelled—a first-class educational system. No small country had better schools, *gymnasia* (high schools), universities, and technical schools than Hungary; and for a country with a population of only 10 million, Hungary also had a remarkably intense artistic and intellectual life.

To take an example from my own family, my father, Vincent Korda, was the youngest of the three sons of a Jewish estate manager who died early, leaving his family impoverished. Nevertheless, Vincent went to a demanding secondary school and *gymnasium* in the provincial town of Kecskemét, where his talent as an artist was quickly noticed; and went on from there to attend the Academy of Applied Arts on a scholarship, and to take classes at the Artists' House in Budapest in 1914–1915. After service on the Transylvanian front in the war, he returned to attend the Academy of Fine Arts in Budapest, and he became a member of the famous artists' colony of Nagybánya, before leaving Horthy's Hungary to study in Italy and in Paris. Though Vincent was poor—he worked as a bricklayer and a scenery painter at the opera to support himself—and Jewish, his education as an artist was paid for by the state; and except for Paris, Hungary then offered perhaps the most advanced and intensive training for artists in all of Europe. Nor was Hungary backward or conservative—avant-garde work was part of the curriculum, and many of the teachers would be among the most famous modern painters of the period between the wars. The same was true for medicine, mathematics, physics, architecture, and chemistry—Hungary produced vastly more educated young people than it could possibly afford to use or employ, particularly after the Allies stripped it of nearly two-thirds of its territory; and since a significant number of these young people were Jewish, they had two good reasons for taking the road to Vienna and beyond.

Hungary also produced more directors, film people, playwrights, photographers, and left-wing journalists than were needed (or desired by the Horthy regime); and they too left to seek their fortune in Paris, London, New York, and Hollywood, where it was soon said, "It is not enough to be a Hungarian; you must also work."

The expatriate Hungarian became, in fact, a mythic figure, particularly in show business, and appeared in a whole series of jokes: "The Hungarian recipe for an omelet begins, 'First steal a dozen eggs.'" "A Hungarian is a man who walks into a revolving door after you and comes out ahead of you." The Hungarian as a shrewd, cunning, and charming émigré became a stock figure, and this image almost obscured the number of Hungarians abroad who didn't become rich and famous, and the number of Hungarians in Hungary who suffered from a woefully feudal rural life and a government that was unapologetically hostile to trade unions and "socialist" ideas about industry. When my uncle Alex—the cosmopolitan movie producer and director Sir Alexander Korda—accepted the first Academy Award ever given for a foreign film (*The Private Life of Henry VIII*) in 1933, he began by saying, "My parents were born poor—but Hungarian!" The audience members roared with laughter because they understood exactly what he meant. To be Hungarian in the world outside the Danube basin was to be, in the eyes of many, a member of a talented and privileged group, so much so that when the Nobel Prize–winning physicist Enrico Fermi was asked if he believed in extraterrestrial beings, he replied, "They are already here—they are called Hungarians!"

This feeling was so widespread that the Hungarians began to lay claim to a truly extraordinary range of historical figures whose genealogy included Hungarian blood, however remote or far back. The Hungarian-American Association, for example, gives as its prize for achievement a bronze medal bearing the likeness of George Washington, who turns out to have had several remote ancestors who were Hungarian, and is therefore "of Hungarian descent."* More than any other ethnic group, Hungarians took pride in listing famous people who were Hungarian, or of Hungarian descent, however remotely, including endless Civil War generals, Nobel Prize winners, table tennis champions, and inventors.

A glance at a recent list of examples of "Hungarian genius" includes, among hundreds of others, the composers Béla Bartók and Zoltán Kodály; the photographers Robert Capa, Brassaï, Martin Munkacszi, and André Kertész; Jozsef Biro, the inventor

*I myself was awarded this medal in 1989.

of the ballpoint pen; the business tycoons Andy Grove and George Soros; the Lauder cosmetics family; the magician Harry Houdini; the actors Peter Lorre, Bela Lugosi, Ernie Kovacs, Leslie Howard, Paul Lukacs, Tony Curtis, and Paul Newman; the actresses Goldie Hawn, Mitzi Gaynor, Jamie Lee Curtis, and the Gabor sisters; the writers Ferenc Molnár, Elie Wiesel, and Arthur Koestler; the nuclear physicists Edward Teller and Leo Szilard; the journalists Joseph Pulitzer, J. Anthony Lukas, and Kati Marton; the comedian Jerry Seinfeld; the rock musician Paul Simon; athletes ranging from Martina Hingis and Monica Seles to Johnny Weissmuller and Joe Namath; the inventor of Rubik's Cube; the African explorer Count Samuel Teleki; Sandor Korosi Csoma ("who walked from Hungary to Tibet, presented the world with the first Tibetan dictionary, and became a Buddhist saint"); the film directors George Cukor, Michael Curtiz, and my uncles Zoltán Korda and Sir Alexander Korda; the clothing designer and entrepreneur Calvin Klein; the architect Lázsló Moholy-Nagy; the mathematicians John von Neumann and Peter Lax; Colonel Agoston Haraszthy, who founded the California wine industry; Max Kiss, who invented Ex-Lax; Charles Fleischmann (of the Fleischmann's Yeast fortune); and Henrietta Szold, who founded Hadassah—not to speak of my father and myself.

Whether Paul Newman, Goldie Hawn, and Jamie Lee Curtis regard themselves as examples of Hungarian genius is hard to say; but it is certain that Hungarians have reached out to claim many of the world's more celebrated people as their own, and that many famous people whose names are not in the least Hungarian proudly claim a Hungarian ancestor on one side of the family or the other. This process did not begin with the period of emigration after World War I and did not end there—Hungarians were doing it as early as the eighteenth century, when Hungarian émigrés tended to be aristocratic adventurers seeking their fortune in the courts of Europe; and in the mid-nineteenth century, when many Hungarians seeking a better life in the New World after the failure of the revolution of 1848–1849 laid claim to aristocratic ancestry and titles that were often imaginary.

To be a Hungarian emigrant is, and has always been, to embark on an adventure; and many Hungarian success stories tend to resemble the plot of a picaresque novel, though Csoma is perhaps the only one to have achieved Buddhist sainthood.

As a result of this process, people have always looked kindly on Hungary, remote as it seemed. The emigrants who left after 1848–1849 were treated as heroes all over the world; and Kossuth himself, the leader of the revolution, was widely admired by everybody, including Queen Victoria, who was not normally an admirer of revolutionaries. This admiration for Hungary—at any rate for Hungary as it was remembered by Hungarians who lived elsewhere—continued throughout the nineteenth century and the early twentieth century. It lapsed briefly in 1918, when the victorious Allied powers took note of the promises they had made to the Czechs, the Romanians, the Serbs, and the Poles during World War I and came to the realization that these claims could be met only at the expense of Hungary, which, as part of the Austro-Hungarian empire, had the bad luck to be on the losing side of the war. This led to the wholesale dismemberment of Hungary; but once that was accomplished, as if by a miracle, the country made its way rapidly back into the good graces of the world.

The Hungarians (with concealed help from the Allies) had overthrown the communist government of Béla Kun (whom Admiral Horthy, with instinctive anti-Semitism, always referred to as "Béla Cohn") by means of a right-wing putsch, and were eager to do business with their former wartime enemies Britain, France, and the United States. Here was at least one small country where communists, and even socialists, were thrown into jail; where there was no nonsense about trade unions or an eight-hour workday; and where the exchange rate was protected by an authoritarian government that smiled on bankers and industrialists, provided they were not Jewish. As if to make up for having amputated much of Hungary's territory (and stranded 1 million Hungarians in hostile countries), the Western powers sent in floods of investment, making Hungary prosperous and burnishing Budapest's reputation as the Paris of eastern Europe.

Despite the exodus of much of its intelligentsia, and its scientific and artistic elite, Hungary, though a fascist country, continued to enjoy a remarkably good press. The Hungarians had paid their debts—or the debts of the Allies—to those of their neighbors who had joined the Allied cause; had rid themselves of a communist dictatorship; and were open for profitable business. Their government enjoyed, among the wealthy and the well connected, a Ruritanian reputation, which it did not deserve but which it worked hard to preserve. Even those who had fled—some in well-justified terror—from the anti-Semitic excesses of Horthy's less cultured supporters nevertheless continued to speak well of Hungary once they reached safety, unlike those who would later flee from Germany in 1933, when Hitler came to power; or in 1938, when he seized Austria.

Hungarians abroad might not speak with one voice—that was hardly possible, given their diversity—but they showed a marked disinclination to criticize their own country. This would play a role in the tragedy to come, for the Russians could hardly have picked a country that seemed to most western Europeans and Americans more harmless and familiar than Hungary to turn into a Stalinist nightmare. Never mind that Hungary had been on the wrong side in two world wars; Hungary seemed, to most people who thought about it at all, part of the West; Hungarians were visible everywhere as celebrities, Nobel laureates, scientists, artists, and filmmakers; and the deliberate (and fraudulent) coup that placed Hungary behind the Iron Curtain and under the thumb of a Stalinist government and its secret police seemed, therefore, a more than usually cynical and brutal example of Soviet policy.

This sympathy for the Hungarians intensified throughout the early years of the cold war, and managed to wipe out, in the minds of most people in the West, the fact that Hungary had fought, with whatever reluctance, on the side of Nazi Germany.

The fact that the Hungarians had a worldwide reputation for being vivacious, gay, creative, charming, and talented merely made it seem to most people all the more sad that they were now trapped behind the Iron Curtain, under the leadership of authoritarian thugs and a Soviet military occupation—victims, apparently in perpetuity, of the cold war.

3.

Hungary and the Cold War

The Hungarians are behaving like Poles. The Poles are behaving like Czechs. And the Czechs are behaving like bastards.

—*Hungarian joke during the Revolution of 1956*

The cold war is a remote memory now, and its early years, in the 1950s, are even more so.

To those of us who grew up during World War II, the cold war never seemed altogether real—for those of us who were British, "the enemy" was forever fixed in our minds as the Germans, who were even then, confusingly, in the process of becoming our allies in the cold war, and whose cars we already admired and coveted. That the Russians were a threat was certainly clear enough; and that they had penetrated farther into Europe than was good for us, or for them, was easy enough to see, even on the more up-to-date maps, on which the "Iron Curtain," to quote Churchill's masterly phrase, was clearly marked with a thick red line. But the Americans took that threat more seriously than most western Europeans—or than the British, who at that time did not yet feel themselves to be Europeans at all.

When my uncle the film producer Sir Alexander Korda married, for the third (and last) time, a beautiful and much younger woman, he bought a house in Kensington Palace Gardens, then as now known as Millionaires' Row, where the embassy of the Soviet Union was inappropriately located. Alex, a truly cosmopolitan man, was always happy to attend cocktail parties at the Soviet embassy, next door to his own house, though he had no sympathy for communism, having experienced it firsthand in Budapest, in 1919, when Béla Kun briefly held power and imposed a communist government on Hungary—which became, briefly, the only western European communist country, to Lenin's surprise and dismay.

During World War II, the Russians had been great fans of Alex's pictures. Winston Churchill had taken a print of *That Hamilton Woman,* starring Laurence Olivier and Vivien Leigh, to Moscow and shown it to Stalin, who had been deeply impressed; and for years it was the only British movie to play in the Soviet Union, where it was hugely popular, although, of course, no royalty payments were forthcoming. Alex was informed from time to time that large amounts of money were available to him in the Soviet Union should he care to come and spend it there, but he had no desire to see Moscow or spend his blocked rubles there, and was content to limit his excursions on Soviet soil to the embassy, where there was an inexhaustible supply of the very best beluga caviar and chilled vodka for a VIP guest like himself.

My aunt Alexa and I often accompanied him to these events. Alexa was a young Canadian woman of Ukrainian descent, and with her high cheekbones, her long blond hair done up in a French twist, and her almond-shaped gray-blue eyes (a touch of Tartar or Kalmuck blood there, perhaps, Alex often teased her, to her great annoyance), she resembled a thinner and more svelte version of the happy collective farmers harvesting wheat who figured in many of the huge socialist realist paintings hung in the embassy—that is, if one could imagine the kolkhoz girls dressed by Balenciaga or Dior, and wearing, say, 200,000 pounds worth of diamonds.

Perhaps because of this, Alexa was much admired by the Russian diplomats. They tended, in those days, to appear almost square—short, broad-shouldered men in baggy gray suits with preposterously wide trousers. They had many gold teeth that they displayed when they smiled, and eyes that hardly smiled at all. They were, after all, hardened communist bureaucrats from a world in which a sense of humor, hand kissing, and a gift for bourgeois charm were not necessarily assets. The contrast between them and their embassy itself was striking—they stood, in their cheap suits and thick-soled Czech shoes, under vast, glittering crystal chandeliers, surrounded by crimson damask walls, gilded carved wood, and monumental marble staircases; behind them was a long buffet table covered by a starched white tablecloth, with whole sturgeons on gilded platters, mountains of caviar in ice carvings of the Kremlin Palace, Georgian champagne in silver coolers, masses of flowers, and many crossed British and Soviet flags.

Many unavoidable toasts were offered in vodka to the friendship of our two nations, and one usually left the embassy with the beginning of a bad headache. I remember the three of us descending the wide white marble steps of the embassy into the graveled courtyard in the early hours of the evening after one such party, as Alex's black Rolls-Royce pulled up (his house was less than 100 yards away, but he considered it an obligation to his belief in capitalism to arrive at the Soviet embassy in a chauffeur-driven Rolls). As we settled into the car—Alex in front next to his chauffeur, Bailey; Alexa and myself in back—Alex lit a cigar and sighed. "If that's the world of the future," he said, "I don't want to live to see it. But you know what they all say now. The optimists in the West are learning Russian. The pessimists are learning Chinese."

In the Royal Air Force, which I was shortly to join—two years of military service was compulsory in those days—the same thought had apparently occurred to the chief of the air staff. The British armed forces had discovered themselves to be almost completely without Russian interpreters and translators at the beginning of the cold war; and now that the Russians had displaced the Germans as the enemy, a joint services task force had been created to force-feed Russian to selected recruits. As somebody who had been educated in Switzerland, knew French and German, and would be going to Oxford on the completion of my military service to study modern languages, I was a natural choice for learning Russian—my name had been placed on the list before I had even packed my bag to take the train up to R.A.F. Padgate, outside Liverpool, and begin my training as a recruit.

In those days, foreign languages still did not come easily, or willingly, to the British. In the British army, Edward Spears had vaulted to high rank and influence in 1914 simply by virtue of what seemed to his contemporaries an astonishing accomplishment: being able to speak fluent French (or at least French that seemed fluent to his fellow Englishmen). This miraculous ability brought him to the rank of general in World War II as Churchill's liaison to the Free French. And this, after all, was the language of our closest ally! Of course before and between the wars British officers commonly learned Urdu, Hindi, Tamil, Arabic, Swahili—whatever languages were necessary to govern the empire—but when it came to the languages of Europe (including France) the general opinion was that all you needed to do to make yourself understood was speak English very slowly and very, very loudly.

The possibility that this might not be the case with the Russians led to the creation of the Joint Services School for Languages in, of all places, Bodmin, Cornwall, to which I was posted as soon as I finished recruit training. There we were woken every morning by Russian marches played on the loudspeaker; were forbidden to speak anything but Russian; performed Russian infantry drill; and in general lived, breathed, and slept Russian. The object, we were informed, was, improbably, to "enable us to pass as Russians" by the completion of the course. To the best of my knowledge, this never happened. Here, however, among the low hills covered with gorse, the pouring rains, and the infinite boredom of Bodmin on a Sunday afternoon, it gradually dawned on us that the cold war was real, and that we were going to play a part in it. Most of our instructors were Russian defectors from communism of one kind or another (including one singularly beautiful young Russian woman, Mrs. Svetlova, married to another instructor, about whom the entire student body, as one man, fantasized every night)—a clutch of mournful Slavs cut adrift in alien Cornwall, within a stone's throw of Jamaica Inn, and determined that we should learn to sing "Bublichki" in chorus and recite Pushkin's poems.

More to the point, what we were doing was top secret—we were forbidden to talk about it to anyone, and our futures were being determined not just by the senior officers of our services but by the occasional mysterious, shadowy visitors in civilian clothes from MI5 (British counterintelligence) or the even more shadowy and mysterious visitors from MI6 (British espionage, the elusive secret service). The people from MI5 looked like policemen or army officers in civvies, with spit-polished toe caps and regimental ties; those from MI6 were distinctly more upper-class, better-tailored types, wearing the ties of the better public schools or of colleges at Oxford and Cambridge. There hovered around MI6 a glamour that Ian Fleming and John le Carré were later to exploit in fiction: in those innocent days, before the defection of Kim Philby, nobody imagined that MI6 was compromised by old-school-tie traitors.

Needless to say, with my natural ability to mimic knowledge of other languages, I did well at Bodmin. It was just like being back in boarding school again, except that I was in uniform and our school was surrounded by barbed wire and had sentries at the gates. I could recite Pushkin's love poems feelingly, and with a perfect accent, to the beautiful Mrs. Svetlova (to no great effect, I regret to say, which was perhaps just as well, since her husband was built like a rugby forward and had been, among other things, a boxing coach in the Red Army), though it was hard to see how the ability to recite Russian poetry would help us win the cold war. Indeed, the joint services—the Royal Navy, the Army, the Royal Marines, and the Royal Air Force—had foreseen this problem. Just in case a knowledge of Russian poetry and folk songs might not prove sufficient to deter the Soviet army from pouring through the Fulda gap into West Germany with a million men, 10,000 tanks, and tactical nuclear weapons, the joint services had provided a formidable regimental sergeant-major of the Irish Guards to keep us up to scratch on the parade ground, and to lead us on night exercises with helmet, full pack, and rifle on Bodmin Moor, where the uncharted bogs frequently swallowed cattle, sheep, and careless drunks taking a shortcut home; or to make us sit in a classroom blindfolded, to strip and reassemble a Bren gun against the clock.

The Russians were coming, we were told again and again—it was a question not of "if" but of "when"—and it was our job to stop them.

Naturally, we were expected to stop them by our ability to translate and interpret Russian—intelligence work, in short—and doubtless only in the last resort would we be expected to stop them with rifle, bayonet, and Bren gun; in the meantime we were left in no doubt that Armageddon was at hand, and that we would have a role to play in it.

Toward the end of my posting to Bodmin I was interviewed by a man in well-tailored tweeds of the kind one might wear for a weekend in the country, an Old Harrovian tie, and suede shoes, who remarked that in addition to Russian my knowledge of Hungarian might turn out to be a bloody useful thing. I pointed out that despite my Hungarian name, I didn't speak a word of Hungarian and knew next to nothing about Hungary. He raised an eyebrow. "Well, it says here that you're proficient in Russian," he said, examining my records. "It's all the same thing, surely?"

Not at all, I explained patiently. Russians were Slavs; their language was Slavonic, distantly related to Polish, Czech, Serbo-Croat, Ukrainian, and so on. But Hungarians were descended from the Mongol followers of Attila the Hun, the Hungarian national hero; Slavs of every and any kind were their natural enemies and they spoke a language that was part of what is called the Finno-Ugric group, comprising Finnish, Hungarian, and Turkish, the only non-Indo-European languages in Europe. While national stereotypes are seldom reliable, ethnic Russians tend to be blond, snub-nosed, and pale-skinned, whereas the Magyar national type is high-cheekboned, hawk-nosed, and dark. Proficiency in Russian would not help me in the slightest in learning Hungarian, which is in any case one of the world's more difficult and impenetrable languages.

"I see," he said, eyes glazing, and dismissed me with a languid wave of the hand. I was conscious of having failed some sort of test, and as I passed his desk I saw that he had written in his black leather notebook, in a neat hand with his fountain pen, "Difficult!" and underlined it twice. I had no doubt that the comment referred to me, not to the Hungarian language, and needless to say whatever secret job having to do with Hungary I was being considered for never materialized.

Instead, I went on to take the R.A.F.'s course for wireless operators; and once I sewed my "sparks" badge up, I was posted to a former Luftwaffe base just outside Hamburg. From there, in conditions of strict secrecy, we flew, by night, in former USAF Super Flying Fortress B-50 bombers (rechristened "Washingtons" by the R.A.F.), up and down the border that separated West Germany and East Germany, provoking the Soviet fighter squadrons to scramble, while we recorded their call signs and tried to identify and place each squadron.

It was tedious work, but since the Russians reacted to our approach like a swarm of bees, there was always the distinct possibility of being shot down by a MIG to keep one on one's toes. It's hard to doze when you can hear in your earphones half a dozen eager night fighter pilots shouting at each other in Russian, only a mile or so away, particularly when one of them says, "Yurik-one, this is Yurik-four, I see the exhausts of the stinking son of a bitch, have I permission to fire?"

Fortunately, the Russians had no more wish to create a lethal aerial border incident than we did, though there were many close calls. On the other hand, our presence in the skies above the border certainly interested them, beyond merely giving their fighter pilots something to do. On one occasion at a nightclub in Hamburg I was surprised when a very attractive woman (of a German type resembling Marlene Dietrich in her early days in Hollywood; or, more recently, Ute Lemper, the cabaret singer at the Bemelmans Bar in the Carlyle Hotel in New York) sat down at my table and spoke to me in Russian. Since I had been staring at her all evening, I was gratified, but also alarmed. I was not in uniform, and there was no reason why she should have supposed I spoke Russian. Much as I wanted to get to know her better, the thought crossed my mind that she might be a Russian agent, or, even worse, a British *agente provocateure* (in every sense of the phrase) testing my loyalty on behalf of MI5 or the Royal Air Force police. Both thoughts put a damper on any possibility of a romantic encounter, and I did my best to pretend that I didn't understand a word of Russian. She then tried again, this time in Hungarian.

That got my attention! Reporting the incident brought me back into contact with MI5, where it was suggested that I might want to pursue the relationship, which they could then monitor; but although this was long before I had read le Carré, it was clear to me that I would be involving myself in a spiderweb of trouble, and that sitting in a four-engine bomber at 25,000 feet while Soviet pilots argued about shooting me down was probably a good deal safer.

On the other hand, the notion that the KGB knew who I was and what I was doing, assumed that I spoke Hungarian, was aware that I knew Russian, and had figured out that I might have a soft spot for blonds with high cheekbones gave me pause, and brought home the reality of the cold war. We (the British) might take it casually (and make fun of the Americans for taking it too seriously), but here, on the ground, in Germany, the cold war took on a whole new meaning. The Russians, certainly, took it very seriously indeed; the Iron Curtain was real; and Soviet-occupied eastern Europe was only a couple of hours away by autobahn from the fleshpots of Hamburg.

The East might be mysterious, but it was never distant.

East Germany (formally, the Deutsche Demokratische Republik, DDR, though not yet recognized as such by the West), Poland, Hungary, Czechoslovakia, the Baltic states, Yugoslavia, Romania, Bulgaria, and even godforsaken Albania were, in those days, all effectively organized as one-party "people's states," controlled by the communist party and the secret police, and pursuing, with varying degrees of success and speed, the elimination of private ownership, the collectivization of agriculture, a strict Marxist-Leninist educational system, complete communist control of the media and censorship of the arts, the dictatorship of the proletariat, the persecution of religion, and unquestioning subservience to the Soviet Union in both domestic and foreign policy.

Each country was led by a kind of homegrown clone of Stalin. Most of these leaders were bullet-headed, heavyset, jowly, scowling authority figures, toughened survivors of Stalin's purges, in front of whom thousands of schoolchildren paraded on national holidays, bearing their leader's portrait, singing his praises, and performing folk dances, followed, for emphasis, by hundreds of tanks. Those not old enough to remember all this need only look at pictures of North Korea today, except that Tiranë, Bucharest, East Berlin, Prague, Budapest, Belgrade, and Warsaw were less colorful and eccentric.

Until his death in 1953, the literal worship of Stalin (symbolized by innumerable huge, brooding statues of him in every shabby, dreary eastern European city) and the need to follow the party line unthinkingly on every subject, however trivial, were strictly enforced by terror, torture, rigged show trials, and the swift execution of those who failed to abase themselves or change their opinions quickly enough. Even after Stalin was gone, his system lumbered on unchanged in the Soviet satellite countries, despite the beginnings of a thaw in the Soviet Union itself. Stalin's gray, pudgy successors, Malenkov, Bulganin, Molotov, and Khrushchev, might find it necessary to repudiate him, cautiously at first, as if he might return from his glass coffin at any moment and have them shot, and to promise "reform"—to allow, as they say in Russian, "a little steam to escape the pot"—but they were not so foolish as to abandon Stalinist methods in the Soviet satellites. Lift the lid off the pot a little in Hungary or Poland, and who could say if you could ever put it back on again?

Apologists for the Soviet Union—of whom there was never a shortage, even among those of us busily spying on its borders—professed to admire its ideal of a classless society, its determination to put socialism into practice, and its sacrifices in World War II; very few people, even the most reactionary of conservatives, actually disliked the Russian people as the Germans had been disliked (for good reason) and by many still were.

All the same, on the other side of the heavily guarded frontier that separated the two Germanies, it was clear that there was a world very different from ours: one of forced confessions under torture for what seemed to be, in Western terms, minor or incomprehensible differences of political opinion or Marxist theory; one in which half the population seemed to be spying on the other half; and one in which a semblance of unanimity was rigorously maintained by huge parades, falsified statistics, the omniscient secret police, and a ruthless show of military power, all of it bringing to mind George Orwell's famous line, "If you want a picture of the future, imagine a boot stamping on a human face forever."

That the Russians had brought this on themselves by embracing Marxist-Leninism and the dictatorship of the proletariat was unfortunate for them, of course, though no doubt their own business. But that they had inflicted their political system brutally in eastern Europe, in their part of Germany, and in the Balkans—and maintained communist rule there, in the final analysis, by the overwhelming presence of Soviet occupation troops and the undisguised threat of using these troops to put down any nationalist or anticommunist uprising—was a different and more dangerous matter. Stifling people's natural desire for independence, freedom, and a decent standard of living is always a risky business.

Countries such as Hungary and Poland had long histories (and even longer memories) and had always thought of themselves as part of western Europe; even Czechoslovakia, created from the debris of the collapse of the Austro-Hungarian empire in 1918, was essentially Western—Prague, after all, had once been the capital of the Holy Roman empire, at the time of the early Hapsburgs.

These countries may have had little in common with each other, but in religion they were predominantly Catholic (with the exception of the Protestant Czechs); they all used the Roman rather than the Cyrillic alphabet; and they had always looked westward toward Vienna, Paris, or London for ideas, fashions, and trade—never eastward toward Moscow.

Russia, on the contrary, even for Slavic peoples like the Poles and the Czechs, let alone militant non-Slavs like the Hungarians, had for centuries been regarded as the home of repression, authoritarianism, and backwardness, a vast landmass of misery, poverty, and ignorance, ruled by a succession of autocratic czars with a fierce appetite for other people's lands, who rejected every modern idea, including, but not limited to, nationalism, republicanism, democracy, the independence of small nations, freedom of the press and the arts, parliamentary government, the rights of man, and almost every other European idea since the Enlightenment and the French Revolution.

Russia, even in the early twentieth century, was a country in which the ideas expressed by Thomas Jefferson in the Declaration of Independence would have seemed deeply subversive, and in which the balance between king and Parliament achieved by the English in the seventeenth century still appeared to the czar revolutionary and dangerous nonsense. Come to that, much of what King John was forced to agree to by his barons at Runnymede in the Magna Carta would probably have appeared totally unacceptable to Nicholas II, had he ever thought to read that document. Nicholas had a great contempt for the tame acceptance by the British royal family of parliamentary government and for the very idea of a constitutional monarchy. This contempt was amply returned by Queen Victoria, Edward VII, and George V, all of whom, though related to the Russian royal family by numerous marriages, were deeply suspicious of the czars' reactionary political views and their concept of themselves as autocrats, as well as by czarist designs on Afghanistan and India. For his part, the czar summed up his feelings, neatly combining two prejudices, by remarking, "All Englishmen are Yids."

To Americans, the cold war was about communism, but to Europeans, even those who were anticommunist, it was, at least subliminally, about Russia as well. "Mother Russia" had always been the enemy of progress; the czar had long been reviled as the "hangman of Europe"; Russia under the czars, whenever it had advanced westward into Europe, had crushed whatever freedom it found and had brought with it the apparatus of Russian absolutism: Cossacks, the knout, censorship, and rule by the army and the police.

Most of Europe had direct experience at one time or another of Russia's presence, in the form of an invading Russian army, even when it was ostensibly an army of liberation. In the early nineteenth century, after the defeat of Napoleon, a Russian army actually reached as far west as Paris, and the impatient call of Russian officers to have their food served to them faster (*Bystro! Bystro!*)—like the varied cold hors d'ouevres with which Russians usually began a meal or of which they sometimes made a whole meal, instead of waiting for it to be cooked from scratch—passed into the French language in the form of *bistro*, meaning a restaurant that served prepared food from a fixed menu so that you didn't have to wait an hour or so for your meal. Culinary innovations apart, those European countries that had experienced the passage of a Russian army had no wish to see the Russians come back again, and the closer they were to Russia itself, the more strongly they felt about that.

Except among European communists, an embattled minority from 1933 to 1945, Russian rule would have been just as unacceptable under the czars as it was under Stalin—indeed the only difference was that in the form of a communist state Russia was more efficient at repression than it had been under the czars, and that it found in every country a certain number of left-wing sympathizers and collaborators. Communism, after all, aspired to be a world movement and appealed to idealism; that had never been the case with czarism. There were communists throughout Europe, not to speak of the United States and Britain; and in every country there were a certain number of people who were sympathetic toward communism without necessarily being party members. By contrast, czarism had been universally feared and despised everywhere outside Russia, even by dyed-in-the-wool reactionaries (and also by a great many people in Russia, of course). Nobody in any country except Russia had ever wanted to be ruled by the czars; but at least some people in most countries wanted to live under communism, though not necessarily in the form of a Russian occupation.

The Hungarians felt even more strongly about this than the Poles. In the mid-nineteenth century, the Poles' attempts at independence (most of what is now Poland was then a part of the Russian empire, and the rest of Poland had belonged since the eighteenth century to the Austro-Hungarian empire and Prussia) had been crushed with a savagery that shocked western Europe, and sent a whole generation of aristocratic Poles into exile. Indeed, the Russian general who crushed the rebellion of 1862 in Warsaw (and sent the czar an infamous telegram that read, "Order reigns once again in Warsaw") was portrayed even in conservative European newspapers as a jackbooted monster strutting over acres of devastated buildings and the piteous corpses of slaughtered women and children, a sword dripping with blood in his hand.

Hungary's experience at the hands of the Russians had been no more benign. In 1848, when revolution swept across Europe, the Hungarians, who chafed under Austrian Hapsburg rule, immediately declared their independence. The Austrian emperor, who already had his hands full with nationalist uprisings everywhere else in the empire (which then included northern Italy, a good part of the Balkans, what became Czechoslovakia, and the southern part of Poland), appealed to his old enemy the czar for help, despite long-standing enmity between the Austrian and the Russian empires; and the czar, who like his predecessors felt it to be his obligation (a kind of imperial noblesse oblige) to support beleaguered reactionary monarchs everywhere, whatever his personal feelings about them, agreed to provide a Russian army to invade Hungary and put down its revolution.

This the Russians did with exemplary efficiency and brutality, defeating the Hungarian army of independence, occupying the country, restoring the Austrian imperial officials to power, and executing the leaders of the uprising. General Bem, one of the leaders of the army, was summarily shot, and became a Hungarian national hero (even though he was in fact Polish); his name would figure largely in the later revolution of 1956. Sándor Petőfi, Hungary's beloved national poet and hero, the Magyar equivalent of Byron or Lermontov, wrote his most famous poems about the revolution, died in battle, and was buried namelessly in a mass grave.

When asked how Austria would ever repay its debt to the czar for restoring Hungary to Hapsburg rule, an Austrian diplomat famously replied, "We shall astonish the world by the depth of our ingratitude." This proved to be true, but more embarrassingly for the future, General Bem and Sándor Petőfi became Hungarian national heroes. Hungarian schoolchildren memorized (and still do memorize) Petőfi's *Nemzeti dal,* the national poem, with its appeal to Hungarians to rise and fight the Russians—not even the communist party could prevent this tradition once it came to power. As for General Bem, placing flowers at the foot of his statue in Budapest would eventually become a way of annoying both the Russians and the Hungarian communist party. Hungary was a country in which history was not part of the dead past but part of the tendentious present and future, and where even statues were potent and troublesome symbols.

After their attempt to achieve independence had been suppressed by Russia, the Hungarians spent the remainder of the nineteenth century exacting from Vienna the best price they could negotiate for acquiescing in their own defeat. The empire itself became the Austro-Hungarian empire or the "dual monarchy," for the emperor of Austria-Hungary was also king of Hungary; and wealthy or aristocratic Hungarians played an increasingly important (and often reactionary) role in government in Vienna, much to the dismay of the emperor, who disliked the Hungarians more than all the other minorities of his dizzyingly multiethnic empire put together.

Hungary was the breadbasket of the empire, however—a vast, still semifeudal source of prosperity, which it was unwise to ignore and impossible to satisfy. Their bid for independence having been crushed by the Russians, the Hungarian upper classes now sought to dominate and control the hated Slavs within the empire, in pursuit of what came to be known as "greater Hungary," which they perceived as including Slovakia, Slovenia, and Croatia, and with a seacoast on the Adriatic.

Since the empire could not survive without cheap food from Hungary, and since the one thing the Emperor Franz Josef could not afford was a repetition of 1848, he was forced to make concessions to the Hungarian landowners and politicians. Much as he disliked nationalism among the ethnic minorities within the empire, he went to the trouble of learning to speak fluent Hungarian—no easy task—and was scrupulous about maintaining his dual roles as emperor of Austria-Hungary and king of Hungary. Every concession he made to the Hungarians was of course at the expense of the Slavs, and deepened the discontent of his Slav subjects, who constituted more than 50 percent of the population of the Austro-Hungarian empire. Fortunately for him, they were split into mutually hostile (and in some cases warring) groups—the Serbs, the Croats, the Bosnians, the Slovenians, the Slovaks, the Czechs, and the Galician Poles, to name but a few—but any hope of reconciling the Slavs to being contented Austro-Hungarian subjects was from the beginning fatally compromised by the need to placate the Hungarians.

In this dilemma may be found, among other things, the seeds of World War I, and the subsequent cataclysmic consequences of the breakup of the Austro-Hungarian empire, many of which are with us still today. It would be exaggerating to argue that the catastrophe was brought about by Hungarian nationalism alone, but Hungarian nationalism certainly played as significant a role as, for example, Serbian nationalism, in putting a match to the powder keg that the Austro-Hungarian empire represented behind the glittering facade of imperial Vienna.

Once, when his advisers were recommending some worthy to the long-suffering Emperor Franz Josef for appointment as a minister, the emperor asked about the man's qualifications for office. "Your Imperial Majesty," he was told, "he is a patriot." The emperor thought about that for a moment and then, with infinite distrust and suspicion, said, "Yes, but is he a patriot for *me*?"

That, of course, was the nub of the problem: there were patriots aplenty in the dual monarchy—Hungarian patriots, Ruthenian patriots, Czech patriots, even Austrian German patriots—but there were hardly any patriots for the idea of a multiethnic empire ruled by the Hapsburgs.

However, one of the striking achievements of czarism in the nineteenth century was to make its royal neighbors and enemies look good by comparison. Franz Josef might be doddering on the edge of senility, his spirit broken by a succession of family tragedies—his younger brother Maximilian had been shot by a Mexican firing squad; his beautiful wife Elizabeth had been assassinated by an anarchist while boarding a lake steamer in Switzerland; his son and heir the archduke Rudolph and his mistress had committed suicide together at Meyerling—but in her lifetime the empress Elizabeth had at least not been in the habit of writing her husband daily, semiliterate, hysterical letters of advice, like those Nicholas II received from Alexandra. In Germany, Kaiser William II might allow one of his officers to dance on a tabletop in pink ballet tights and a tutu for after-dinner entertainment, but at least his policies were not influenced by the likes of Rasputin. Compared with the Russian court, the courts of Berlin and Vienna seemed almost benign, moderate, and sane.

Hungarians took note of this. Most of them did not want the breakup of the dual monarchy; they merely wanted more power, autonomy, and influence within it, as well as the right to bully and dominate the adjacent Slav minorities in pursuit of "greater Hungary." All this was in line with economic development in the late nineteenth century. Eastern Hungary was being rapidly industrialized; mines were opening up to exploit Hungary's mineral wealth; capitalism was beginning to alter what had hitherto been predominantly a rural, feudal, agricultural economy, with nothing much in the way of an educated middle class. Hungary might have a reputation for wildness, "Gypsy" music, rich spicy cooking, and beautiful women; wolves and bears might still present a danger to travelers in Transylvania;

wealthy foreign visitors might still be entranced by shooting parties on huge feudal estates where time seemed to have stood still—but in fact the Hungarians were progressing rapidly toward the twentieth century. Hungary might have become part of a revitalized Austro-Hungarian empire, like the Czechs, had not the empire itself under Franz Josef been impervious to political reform and so deeply mired in the unpredictable violence of Balkan politics on its vulnerable southern flank that it was at the mercy of events.

These events, of course, are too well known to require repetition here, except to note that following the assassination in 1914 of the heir to the throne, the singularly unlikable Archduke Franz Ferdinand, and his morganatic wife while they were on an official visit to Sarajevo, everybody's worst nightmare came true. Pessimists had always foreseen that Austria-Hungary would lose control of events in the Balkans and draw the rest of Europe into the abyss, and so it proved. As an integral part of the empire, Hungarians found themselves drawn into war first with Serbia; then Russia; then the Western allies, France and the United Kingdom (the very countries Hungarians admired most); and finally Romania, Italy, the United States, and even, improbably, Japan.

Defeat, when it came, brought with it a whole new and unrecognizable game, in which for a time it seemed as if Hungary, after much sacrifice, might have drawn a winning card. The Austro-Hungarian empire was gone at last, shattered into a new world of small, independent, and mutually hostile countries. Imperial Vienna was now merely the capital city—the size of Paris—only of German Austria, a tiny country the size of Switzerland, while new countries with uncertain frontiers, Poland and Czechoslovakia among them, had emerged from the ruins of the empire. The Russian empire had vanished too, and Russia, mired in revolution and civil war, had retreated hundreds of miles back from its previous frontiers, for the moment a threat to nobody but itself.

As a small nation craving independence, Hungary had high hopes of being granted reasonable peace terms by the Allies, particularly since there were a substantial number of Hungarian immigrants in the United States. These hopes, alas, were not to be realized. The Hungarian Parliament selected a liberal landowner, Count Mihály Károlyi, an idealist, to head the government, in the hope that he and President Woodrow Wilson would hit it off, but that was not to be. The Allies had obligations to Poland, Czechoslovakia, and Romania, and intended to honor them at the expense of Hungary. Polish troops had actually fought on the western front, and in Ignacy Paderewski, the world-famous pianist, the Poles had produced a political figure whom everybody recognized and admired. The Czechs too, in Edvard Beneš, had a leader who was greatly admired in the United States. Even Romania, although it had entered the war on the Allied side late (and with disastrous consequences for itself), had, in Queen Marie, a glamorous and impassioned celebrity as a spokeswoman for her nation's claims, most of them against Hungary. Hungary had fought to the end—without enthusiasm, it is true, but with exemplary loyalty to the doomed Austro-Hungarian empire—and in the age of Wilson's Fourteen Points and of the self-determination of nations, its claims to large amounts of land mostly populated by Slavs and non-Magyars won it no friends at Versailles, where the new frontiers of Europe were being drawn.

As the Hungarians were shortly to discover (to paraphrase Orwell), seen from the Paris Peace Conference all forms of nationalism were equal, but some were more equal than others. Wilson and Clemenceau looked with favor on Polish nationalism, based on a reconstituted Poland, formed out of what had been Russian, German, and Austro-Hungarian territory; they also favored Czech nationalism, even though it involved the creation of a country formed out of a shotgun marriage between the liberal Protestant Czechs and the reluctant and reactionary Catholic Slovaks, and included what soon came to be seen as a dangerous number of Sudeten Germans. But Hungarian nationalism held no appeal to the leaders of the Allied powers in Versailles. Hungary seemed to them feudal, backward, determined to hold on to land that didn't (or shouldn't) belong to it, and insufficiently apologetic for having fought on the wrong side for four years.

Nor did Hungary present them with a picture of stability, despite Károlyi's best efforts to put a good face on things. Inflation soared, apparently beyond control; aristocratic grandees, whose titles derived from what was now an extinct empire, plotted right-wing coups; Hungarian prisoners of war returning from Russia brought communism back with them; a variety of Hapsburg pretenders were on Hungarian soil—Hungary, independent at last, seemed to have nowhere to go but down.

Beset by unrest on all sides and the threat of coups and plots from the left and the right, Károlyi struggled on, staying barely afloat in the murky waters of Hungarian politics until 1919, when an Allied delegation presented the Hungarian government with a map of Hungary's new frontiers, causing the Hungarian deputy foreign minister to faint at the sight of it. Hungary was reduced to a fraction of its former size; huge chunks of territory over which Hungary had ruled undisputedly for years were lopped off to satisfy the Poles, the Czechs, and the Romanians; hundreds of thousands of Hungarians would wake up to find themselves overnight living in foreign countries, where their neighbors and the government hated them. This nightmarish scenario, intensified by the threat of invasion by the Romanian army to enforce Romania's territorial demands, brought down Károlyi's government and plunged the country into chaos, in which the only people willing to seize power were the communists.

Thus Béla Kun came to power, in January 1919; Hungary was transformed, to the horror of the middle and upper classes—and the Allies—into the Hungarian Soviet Republic, a full-fledged dictatorship of the proletariat on the Danube. Kun, though his reputation was much blackened in the years that followed, was a Marxist theoretician rather than a man of action, and his method of government relied on persuasion and propaganda rather than violence. It did not help Kun that in a country where anti-Semitism was widespread he was a Jew. Despite Kun's many appeals to Moscow for help, Lenin had problems of his own and in any case was not inclined to see Hungary as the right country in which to sponsor the westward expansion of the communist revolution. All Marxist ideology taught that Germany was the country in which this must take place, and Germany in 1919 seemed on the brink of revolution in any case. Kun received from Moscow any number of crisply worded telegrams urging him to take ruthless action against the class enemy, and a few Hungarian-speaking envoys bearing the same message—but he received no arms, no money, no offer of armed assistance. He was on the verge of succumbing to a joint invasion of Allied and Romanian troops when, in August 1919, a right-wing counterrevolutionary army under Admiral Miklós Horthy marched into Budapest, disposed of the communists with swift brutality, checked the advance of the Romanians, and instituted the first fascist government in Europe—one which was to last twenty-six years.

Those of Kun's followers who escaped took refuge in the Soviet Union, where Stalin had most of them quietly murdered during the purges of the 1930s. Hungary itself settled into a long, sleepy period under its new leader, who declared himself regent of the kingdom of Hungary, though with no intention of inviting any of the prospective kings to reassume the crown of St. Stephen. Horthy was a conservative, even a reactionary, an anti-Semite, opposed to trade unions and almost everything "modern"; but once the communists and the invading Romanians had been dealt with, he was all for the quiet life.

It was apparent to him, as it was to most Hungarians, that caught as their small country was between Germany and the Soviet Union, survival would depend on subtle, deft diplomacy—something on which Hungarians optimistically prided themselves. Of the two powers, Horthy naturally preferred Germany—he had been, after all, an admiral in the German-speaking navy of Austria-Hungary; and he had no sympathy for communists, or even for mild socialists. But in his own cautious way he was determined to get the best price he could for Hungary's friendship, starting with the return of Transylvania, and then of the lands "stolen" by the Poles, the Romanians, and the Czechs.

Horthy was not a young man, and his style was not that of his imitators in Italy and Germany. He did not posture and strike belligerent poses like Mussolini; nor did he rant and rave and threaten, as Hitler would. The huge open-air mass rallies beloved by Mussolini and Hitler did not stir Horthy's blood. He had ridden into Budapest on a white horse, in his admiral's uniform, in 1919 (thus re-creating the legend that what the people were looking for was the arrival of a "man on a white horse"—i.e., a military dictator—begun with General Boulanger's failed coup in late-nineteenth-century France). But at heart Horthy preferred the comfort of a Mercedes-Benz limousine—he was no swashbuckler but rather a canny, cautious, conservative politician. It is no accident that he ended up spending his last years, after World War II, in the comfort of a villa in Estoril, Portugal, surrounded by former kings as neighbors, rather than ending like his fellow fascist dictators, cremated in the garden of the *Führerbunker* in Berlin like Hitler and Eva Braun, or half-naked and strung up by his heels upside down by the mob like Mussolini and his mistress Clara Petacci.

Horthy was, in any case, a family man, devoted to his wife, and like Emperor Franz Josef, in whose navy he had served for so many years, he was equally devoted to decorum, discipline, and the bourgeois comforts of life. He resembled in many ways an archetypal figure of nineteenth-century and early-twentieth-century central Europe: the stern, authoritarian paterfamilias, affectionate, but distant, demanding, and intolerant toward dissent of any kind—exactly the kind of father figure, in fact, who played such a potent role in the neuroses that Sigmund Freud's patients brought to him, and to whom Freud himself, in private life, bore a certain resemblance.

The fact that Hungarian fascism was less hysterical than fascism in Italy or Germany—or that it put forth, for many years, a blander, more comfortable image and a leader who seemed profoundly respectable, and even dull, rather than a rabble-rousing fanatic—does not, of course, make it any less fascist.

Before World War II—indeed almost until 1944—Hungarian anti-Semitism was a good deal milder than German anti-

Semitism under the Nazis, but this is not to excuse it, or deny its existence. Laws defining who was Jewish and who was not were based on religion rather than race, thus giving a good deal more scope for "conversion." Also, the Hungarian Jews themselves were sharply divided between the more "assimilated," middle-class, urban Jewish population and a rural population in which clusters of Orthodox or even Hasidic Jews stood out sharply from the surrounding Magyars.

Through the 1920s and early 1930s, Hungary managed to survive better than many had anticipated as an independent country, no longer part of a multinational empire, and reduced to a fraction of its former size. Agricultural richness; a wealth of natural resources; a hardworking, well-educated, resourceful population; and a growing industrial base produced a modest level of prosperity. Hungary's relations with its immediate neighbors were strained, but peaceful. To the west, Germany was moving like a sleepwalker from crisis to crisis; to the east, Russia was moving from revolution and civil war into the deadly stability of Stalinism; and while it is never reassuring for a small country to be sandwiched between two giants, at least Hungary was not one that either of its two giant neighbors dreamed of destroying and annexing, as, for example, Poland was. Even the beleaguered governments of Weimar Germany had never reconciled themselves to the "Polish Corridor" or the loss of Danzig; nor had the ordinary German—and Russia, whether communist or not, never gave up the hope of regaining all or most of Poland. Nobody thirsted for Hungarian land—on the contrary, it was the Hungarians who never lost sight of regaining what they had lost to Romania, Czechoslovakia, and Poland, and waited for the opportunity to do so.

A spirit of mild friendship existed in principle between the

Italian and the Hungarian fascist regimes. Like Mussolini, but on a smaller scale, Horthy was treated with wary respect by the Western democracies—the fact that he was against communism, socialism, and trade unions was in his favor among "respectable" people, of course; and at the time anti-Semitism as such was unlikely to offend most British, French, and American politicians and bankers—not to speak of the Vatican—provided it was discreet and did not involve unpleasant displays or violence in public. In Rome, Hungary was thought of as a "client state," with possibly conflicting ambitions in the Balkans, where the Italians had seized Albania and had great ambitions to expand farther at the expense of Yugoslavia and Greece. Still, Hungary and Albania were the only countries not to vote against Italy in the League of Nations over oil sanctions to punish Mussolini for attacking Abyssinia; and there remained to the end a Rome-Budapest axis, more important to Horthy than it was to Mussolini.

"Fascism" is of course used today as a loose term of political opprobrium, but in its own day, and in its original form, it had a more specific meaning. On the deeper level it involved an understanding between "people of property" (i.e., big business, the right-wing press, and the banks), the church, and the middle classes, usually under the leadership of a single charismatic leader, to support nationalist, reactionary groups of veterans who had fought in World War I, in an effort to prevent the spread of communism, undermine trade unions, and bypass parliamentary democratic government.

Since the police were likely to be sympathetic toward veterans (or were veterans themselves, if only of the military police), they, as well as the army, which was officered by the middle and upper classes, could usually be relied on to support fascism, tacitly or openly. A modest degree of force—the rubber truncheon; the castor oil treatment; the occasional outright murder, such as that of the Italian journalist Matteotti—was applied when necessary, but the general object of fascism was to stifle dissent and bolster the existing establishment, while producing enough drama in the way of rallies, parades, propaganda, and the occasional foreign adventure to siphon off the energy of the lower middle class and the working class, who might otherwise have turned toward radical social reform or worse, communism.

Seen from Rome or Budapest, Nazism before 1933 looked similar (hence the comforting belief of Mussolini and Horthy that Hitler was their "pupil"), but the impulse behind it was vastly more evil, bolder, and more destructive. Hitler was not the servant of big business, the army, or the bourgeoisie, despite what people still like to believe. His nihilism; his restless urge for change; his willingness, even eagerness, to condone and order bloodshed on a large scale—indeed the whole messianic side of his nature—made the Nazi movement infinitely more dangerous than fascism. *Weltmacht oder Niedergang*, "world power or destruction," was not a motto that would have appealed to Mussolini or Horthy; but when Hitler said it, he meant exactly that, and he would end his life following its ultimate logic.

The fascist world was of course turned on its ear in January 1933, when Hitler unexpectedly became Germany's chancellor. Germany, even in defeat, was a great European power, so that a fascist Germany was therefore an altogether different kind of beast from a fascist Italy or Hungary; and Nazism was an altogether more radical movement, which included some very strange and threatening ideas indeed. Horthy and Mussolini had made their peace with the church and maintained cordial relations with the

Vatican; Hitler intended to subvert and uproot religion from German life, and replace it with the cult of Nazism, in which he himself was the messiah. Horthy and Mussolini had territorial claims, which, with patience, could probably have been met; Hitler dreamed of "world power" and of a Germany that stretched from the Rhine to the Urals (and also dreamed of exterminating those peoples he didn't want in it, starting with the Jews, of course, and eventually eliminating the Slavs).

The air over Germany from 1933 to 1945 was thick with a heady mixture of energy, cruelty, and extravagant ambition, as well as the residue of a century of pernicious Germanic nonsense, which included drastic solutions to social problems ranging from forced euthanasia for the mentally and physically handicapped to the breeding and perfection of an Aryan super-race, and strange theories about everything from Nordic runes (with which Himmler hoped to prove that the Japanese were Aryans) to the head shape of "the typical repulsive Jewish commissar type." In pursuit of this head shape, SS squads would eventually roam Poland and the Soviet Union, equipped with exact instructions on how to kill their victims without damaging the head, and special containers in which to send the head samples back to the anthropologists of the SS Institute of Race Research.

None of this was at first evident—the coming to power of Hitler seemed at first, in fact, to provide Hungary with a powerful and ideologically sympathetic ally, and, as Germany rearmed, a strong and growing market for Hungarian raw materials and manufacturing. Hitler kept a signed photograph of Admiral Horthy in a silver frame conspicuously displayed on his desk, in honor of Horthy's role as the first fascist in Europe, and professed great admiration for the regent, as Horthy was known.

However, despite the fact that the Führer had been hastily naturalized as a German citizen when it was realized that this small step had been overlooked on his way to power, he remained in many ways an Austrian, by birth, by upbringing, and in his general attitude toward central Europe.

Hungary might seem distant and faintly romantic to Prussians, but Budapest was less than 250 miles from Vienna, and until 1918 had been part of the same empire. Hitler had no romantic illusions about Hungarians, and he instinctively distrusted anybody who spoke what Austrians called *Miklós-deutsch*—that is, German spoken with a distinctive Hungarian accent. He was unmoved by the czardas, by Gypsy music (he intended to exterminate the Gypsies down to the last man, woman, and child in any case), or by Hungarian charm. As an Austrian, he held the Hungarians responsible for many of the terminal problems of the Austro-Hungarian empire, and he did not wish to be dragged into their disputes with their neighbors, or have his Balkan policy determined by Hungary's interests. If the Hungarians wanted his help in reclaiming the lands they had lost at Versailles, they would have to dance to his tune—"He who wants to eat must do his share of the cooking," the Führer said, with his usual blend of brutality and cold realism.

Hungarian charm, polish, and duplicity might work wonders in Rome, and even, to a degree, at the League of Nations in Geneva; but they had the opposite effect in Berlin, where heel clicking, hand kissing, and *Schwämmerei* were being replaced by altogether more brutal manners, and where anti-Semitism was swiftly moving beyond merely daubing insulting slogans on synagogues, setting quotas in universities and medical schools, or taunting the occasional bearded rabbi in the street.

Increasingly, Germany's few political allies came to feel that they were like passengers on a runaway train. There was no question of getting off the train—for a small central European country like Hungary, surrounded by enemies, the protection offered by friendship with Nazi Germany was indispensable—and at the same time there was no way to slow the train down. Czechoslovakia had a military pact with France and the support of Great Britain (neither of which would help it a bit in 1938); and Poland constituted an integral part of the French defense strategy, the idea being that the Czechs and the Poles would act as a cordon sanitaire against the westward expansion of Russian communism, and in the event of war with Germany would keep the Wehrmacht busy on the eastern front while the French army hunkered down behind the Maginot Line and waited on events. But Hungary was not a beneficiary or a creation but rather a victim of Versailles—along with Germany and Turkey, it had been dismembered, laden with war guilt, and severely punished by the victors.

Much as the Hungarians liked and admired France and the United Kingdom, they were not naive enough to suppose that they would get much help from Paris or London once Russia regained strength. Horthy was many things, but he was not naive, and as the 1930s went on, one disastrous event after another confirmed his view that friendship with Germany was the only ace in his hand. Hitler reoccupied the Rhineland; the Allies and the League of Nations did nothing. When Hitler rearmed and started to build the Luftwaffe, they still did nothing. When he violated one of the major clauses in the Treaty of Versailles by his *Anschluss* with Austria (thus advancing the border of Nazi Germany to the Hungarian frontier, and effectively isolating the Czechs), again they did nothing. When he threatened to attack and dismember Czechoslovakia on behalf of the Sudeten German minority, Daladier and Chamberlain flew to Munich and agreed to cripple the state the French and the British had created less than twenty years earlier; and when they returned home they were cheered by crowds of their fellow countrymen for doing so.

Seen from Budapest, these events spoke for themselves. If the Allied powers were unwilling or unable to support a democratic state which they had themselves created, and which played, at least in theory, a major role in their defense, what kind of help could Hungary expect? The answer was, clearly, none.

Not that there were any tears for Czechoslovakia in Hungary. Early on in the crisis, the Hungarians had presented their claim for the return of Slovakia to Hungarian rule once the Czech state collapsed; Hitler listened with what the Hungarians supposed was sympathy, and encouraged them to put additional pressure on the Czechs. When push came to shove, however, Hitler, having occupied Prague, was persuaded to turn Slovakia into a German client state with a fascist government and a Catholic priest, Monsignor Tiso, as the premier. The Hungarians had missed the boat—their army had been in no shape to threaten the Czechs or occupy Slovakia, and Hitler saw no reason to reward them for inaction, or for agreeing to what they could not in any case prevent. His mind was now set on crushing Poland, which would be a much safer move if the Soviet Union shared in the spoils. He still hoped to avoid a general war, but if it happened, Romanian oil would be absolutely necessary to keep the Wehrmacht moving. Germany's friendship with the Soviet Union and Romania—the very countries the Hungarian government feared and disliked most—would mean that Hungary could not expect to get more than a few crumbs off the table with regard to reclaiming its territory. This no doubt explains, as much as the poorly equipped state of the Hungarian army, why Hungary remained neutral from the outbreak of World War II in September 1939 to the German attack on the Soviet Union in June 1941.

Neutrality, as everybody knows, has its advantages, but Hungary was not Switzerland. Planted in the middle of central Europe, with a government that sought friendship (and favors) from Germany, and a growing and increasingly threatening political minority that imitated the Nazis, Hungary was in no position to sit the war out, nor was Hitler disposed to let it do so. Diplomatic and economic pressure was applied, accompanied by demands that Hungary should take a "harder" position against the Jews and bring its anti-Semitic laws—and more important, the severity with which they were carried out—more into line with Germany's own Nuremberg laws.

In a spirit of fatalism that bordered on the suicidal, Hungary joined the Germans in their war against the Soviet Union, in the expectation that it would be over quickly—as indeed everybody supposed it would be—and that Hungary would share modestly in the spoils. When the German army ground to a frozen halt within sight of Moscow in the winter of 1941, it became clear to the Hungarians—now that it was too late—that they had lost their bet. The Soviet Union did not collapse, as Hitler had foreseen ("The whole rotten mess will collapse once we kick the door in," he had promised); instead, the war in Russia would be long, bloody, and destructive.

The Hungarian army, ill-equipped for a long campaign or for the Russian winter, something less than enthusiastic, and poorly supported by the Wehrmacht, suffered heavy casualties, while at home the Horthy regime found itself under increasing pressure from Germany to harshen measures against the Jews. In December 1941, the unthinkable happened—Hungary was forced to declare war on the United States,* and the United Kingdom declared war on Hungary. Given the fact that there was a large Hungarian-American population in the United States, and that Great Britain and the United States were the two countries Hungarians most admired, this was a fatal diplomatic step, made worse by the need to placate the Germans as the Hungarian army was pushed back and decimated by superior Soviet forces. Jews were rounded up and sent as forced labor to dig antitank ditches on the Russian front, where they died in large numbers; communists, socialists, and trade unionists were arrested and imprisoned. Still the Germans were not satisfied, particularly when they began, inevitably, to hear rumors that Horthy's government was secretly trying to negotiate its way out of the war.

By the early spring of 1944, the Hungarians were getting desperate—the Hungarian army was in tatters, the Red Army was advancing closer and closer to Hungary, and the Allies had still not landed in France. At the end of his patience with Hungary, Hitler browbeat Admiral Horthy into appointing a more pro-German government, and sent eight German divisions into Hungary to ensure Hungarian compliance. They were shortly followed by SS Obersturmbannführer Adolf Eichmann himself, to oversee the beginning of measures to implement the

* The Hungarian minister in Washington, to his great embarrassment, was obliged to read the Hungarian declaration of war to an incredulous Cordell Hull. "Has Hungary any claims or grievances against the United States?" the secretary of state asked. "No, no, of course not," the Hungarian diplomat replied soothingly. "Against whom does Hungary have claims or grievances then?" A shrug: "Romania, Your Excellency." "And is Hungary declaring war on Romania?" The Hungarian sighed and held up his hands helplessly. "No," he said sadly; "unfortunately the Romanians are our allies."

"final solution" of the "Jewish problem" in Hungary. The first transports of Hungarian Jews to Auschwitz took place, and these deportations would not end until the SS had destroyed Auschwitz itself, by which time nearly 500,000 Hungarian Jews had been murdered there.

The Allies' landings in Normandy in June 1944, followed by the attempt against Hitler's life in July and the liberation of Paris in August, convinced Horthy that he must act quickly if Hungary was to join the winning side at the last moment. In this regard, as in so many others, Horthy's instinct was sound, but his timing was fatally flawed. The Allied armies, bogged down by transportation difficulties and limited supplies of fuel, failed to take Antwerp, which would have provided them with the major port they needed close to their front line; they also failed to reach the Rhine, let alone cross it, and the German army, resourceful as always, managed to end a retreat in the west that was turning into a rout and form a strong defensive line. The war would not, as many had predicted, be over before Christmas.

The Hungarians' hope for a quick victory by the Allies in 1944 was as futile as their hope for a quick victory by Germany against the Soviet Union in 1941 had been. This time the consequences were even more tragic. The war would go on until May 1945, and Hungary would be squeezed between the infuriated Germans and the remorselessly advancing Russians. The Germans were determined to fight the Red Army for every square foot of Hungarian territory, bringing about widespread destruction as a kind of punishment for betrayal and lack of enthusiasm—and of course to keep the murder of the Hungarian Jews going for as long as possible. Budapest, which, like Prague, had been spared by the Allies' bombers, would be utterly destroyed by the Germans' determined stand there and the Soviet siege of the city. More than 25,000 civilians would be killed, and a quarter of the city's houses damaged or destroyed. A Swiss diplomat reported to Bern, "Half the city at a rough estimate is in ruins. Certain quarters have . . . suffered more than Stalingrad. The quays along the Danube, and in particular the Elisabeth Bridge and the Chain Bridge, are utterly destroyed. . . . The Royal Palace has been burned to the ground. The Ritz, Hungaria, Carlton, Vadazkurt, and Gellért hotels are in ruins." *

Amid the ruins and the chaos of street fighting and artillery bombardment, the German SS and paramilitary units of the Hungarian Nazi Arrow Cross party—really no more than armed, uniformed thugs and criminals—sought out Jews and murdered them, while from abroad Franklin Roosevelt, the pope, and the British government begged the Hungarian government to put an end to the deportations and murders. Horthy, who would gladly have agreed to do so at this point, was powerless. His beloved son Nicholas was kidnapped by the Gestapo and smuggled out of Hungary rolled up in a carpet, as a hostage for his father's good behavior; and Horthy himself would shortly be arrested on Hitler's orders and imprisoned in Germany, replaced by an Arrow Cross government that was determined to fight on to the bitter end.

The end, when it finally came, was tragic and ignominious. Its capital in ruins, its countryside ravaged by war, its people starving, Hungary, in April 1945, exchanged a German occupation for a Soviet one. Hungary's attempt to surrender to the Western Allies failed completely, and although the Iron Curtain

*It is interesting to note the typically Swiss concern for hotels. Not for nothing are the Swiss known as a nation of innkeepers.

had not yet come down, Hungary would shortly find itself on the wrong side of it. Political power had evaporated in Budapest, where there was no government left with which anybody would have willingly negotiated.

Not that it mattered, for all practical purposes. In the provincial city of Debrecen, under the close supervision of the Russians, a "provisional government" had already been formed.

4.

Salami Tactics

For the second time in the twentieth century, Hungary had chosen—or been obligated—to fight on the losing side in a world war. The first time cost it two-thirds of its territory, and left millions of Hungarians living under hostile governments, with neighbors who hated them; the second cut it off from the West and left it to the tender mercies of Stalin and the Hungarian henchmen he had not killed in the great purges.

This was not only a political and human disaster, but also an economic one. To the west of what would soon come to be called the Iron Curtain, the countries of Europe would receive aid and investment in vast quantity from America to put them back on their feet again, as well as the blessings of the Marshall Plan. Even Germany, the cause of everybody's misfortune, would be rebuilt with American money—the western part of it, anyway—while just across the border from Hungary, Austria, birthplace of many of the most notorious Nazis, including the Führer himself, would actually be treated as a "victim" that had been "occupied" by Nazi Germany, rather than as an integral part of it.

The Russians, far from subscribing to the American notion of rebuilding those countries they now occupied, pillaged and looted them instead, sending vast amounts of industrial machinery and rolling stock back to the Soviet Union. Trade with the East was, in any case, a chimera—the currencies of the eastern European countries had little real value and the Soviet ruble still less; nor did the Russians have anything to sell their new client states. Poland was a wasteland full of horrors, and Hungary hardly better off—only Czechoslovakia, which had escaped the fighting, was still a functioning industrial power.

What the Soviet Union *did* have to export was ideology, and the wherewithal and experience with which to organize and run a police state controlled by a single political party. And so long as Stalin was alive, the will to impose communism on other nations was never lacking; nor was there any hesitation in using the methods that had proved so productive in the Soviet Union: the midnight arrest, torture, the forced confession, the show trial, and the inevitable execution. As for the Hungarian communists who returned to their homeland in the wake of the Soviet army, they were a hardened lot, very often survivors themselves of terror, arrest, torture, and prison camps in Russia, and with an endless list of old grudges to pay off and scores to even once they were at home again and in power. Some of them, in 1956, would be "rehabilitated" and treated as heroes, posthumously or not, but none of them—not even Imre Nagy, who would become the martyred leader of the revolution—was without blood on his hands. Those few foreign communists who had survived Stalin's terror were of necessity unsentimental realists who understood the meaning of power.

Still, Stalin moved cautiously at first. He felt he had a valid title to the countries of eastern Europe—indeed, in October

1944, after a long dinner in the Kremlin, Winston Churchill had written down in pencil on a half-sheet of paper the relative percentages of Russian and British influence that should be maintained in these countries once the war was won.[1] This casual (and cynical) approach to the future of Europe, which would have caused lively alarm in Franklin D.Roosevelt and the U.S. State Department, resulted in Stalin's getting 90 percent control in Romania, Britain's getting 90 percent in Greece, and Hungary's being split fifty-fifty between them.* While puffing contentedly on his pipe, Stalin obligingly initialed the paper with his blue pencil (he used a soft red pencil for initialing his approval of the names of people he wanted executed), and allowed Churchill to keep the document, which then had to be concealed from the Americans lest they be shocked by this example of old-fashioned, pre-Wilsonian great-power diplomacy, about which they had not been consulted. In the event, of course, as Stalin doubtless anticipated, the Soviet Union would have 100 percent of the power wherever its soldiers reached, and that was that.**

On the other hand, Stalin felt that the longer his allies could cling to the illusion that some degree of democracy would be put in place in eastern Europe (and that they would have some degree of influence over what happened there), the better for all concerned. The Soviet Union was exhausted and impoverished

*Eventually Stalin took 100 percent of Romania, tried to take more than his share of Greece by means of a communist uprising that was put down by British troops, and simply ignored attempts on the part of Great Britain and the United States to influence events in Hungary. (He did, eventually, live up to his agreement with Churchill about Greece, however.)

**The one exception was Austria, from which all the Allies eventually withdrew together by mutual agreement.

by the war, and despite feverish efforts four years would elapse before it tested a nuclear weapon, without which any confrontation with America would be a risky business—Stalin was, as always, willing to take the long view. The fruit would fall into his hands when it was ripe, as he was fond of saying.

In the meantime, a government of sorts was emerging in Hungary, in Debrecen rather than Budapest, which was in ruins and still full of sinister violence. The sudden disappearance of Raoul Wallenberg, the Swedish diplomat and humanitarian who had risked his life over and over again to save Jews from the SS, and who was spirited away secretly by the Russians to vanish into the gulag, where he died or was killed, was a perfect symbol for the way the Soviet Union intended to deal with people who were inconvenient in "liberated" Hungary—they disappeared, and were never heard from again. The government at Debrecen was designed to plaster this unpleasing reality over with at least a facade of democracy. A provisional "national assembly" appointed an equally notional "government," led by, of all people, a tame aristocratic former Horthy general, and containing representatives from most of the "bourgeois" political parties—an artfully contrived group in which there were only two outright communists visible among eleven ministers—to negotiate a peace treaty. The provisional government was deftly created and stage-managed by the Russians, chiefly for the purpose of keeping the United States, Great Britain, and France quiet.

This was the kind of thing at which Stalin and the men around him, particularly Mikoyan, Molotov, and Marshal Voroshilov, were past masters—the puppets were moved so skillfully that hardly anybody even noticed the strings. The men who had served in Horthy's government and remained comparatively undisgraced, the leaders of the bourgeois smallholders and peasant parties and the leftist but noncommunist social democrats, moved through the motions of liberal, democratic government, while the real power in the country lay increasingly in the hands of the Soviet ambassador, the occupying Red Army, the NKVD (the Soviet secret police), and the Hungarian communist party. A peace treaty was drawn up and signed; Hungary agreed to pay huge reparations to the Soviet Union—a bill that was still coming due at regular intervals in 1956—as well as to pay for the cost of the Russian occupation; the big estates were systematically broken up; and efforts were made to form collective farms. An election was held, in which communists won only 17 percent of the vote, despite great efforts on the part of the Russians to rig the election on their behalf; but this disappointing showing did not prevent the communists from ending up with more seats in the cabinet than the winning party after Marshal Voroshilov, the president of the "Allied" Control Commission, made it clear that he would not accept any other outcome.

The Hungarian communist party was sharply split between those who had remained "underground" in Hungary in great danger through the war and those who had fled to Moscow before the war. It almost goes without saying that those who had fled to Moscow—and survived the purges—were more trusted by the Russians than those who had stayed behind, however courageous they had been. Chief among the former was the brutal, bullet-headed communist leader Mátyás Rákosi, whom the Horthy regime had handed over to the Russians in exchange for the banners taken from Kossuth's defeated army by the czarist army in 1849—further proof of the extraordinary power of historical symbols in Hungarian politics. In Moscow, Rákosi had learned the importance of patience and the need to follow the party line diligently—under Stalin, those who neglected to do so, or failed to predict where it would lead, paid for their inattention with their life in Moscow. Rákosi also learned the value of terror as a political instrument. The members of the Hungarian communist party walked a tightrope—every word they uttered or wrote was carefully scrutinized, and the slightest deviation from the party line led inexorably to arrest, torture, disgrace, or execution. There were no trivial matters, no exceptions—the wrong opinion about a novel, or even a review of a novel, could have consequences as fatal as expressing the wrong political ideas in a newspaper article or in a speech. There was a "correct" communist viewpoint on everything, however minor it might seem, and party members were expected to be as alert to the swift-changing party line in Moscow as animals on the African plains are to the approach of a predator.

The first great schism in the postwar communist world was Tito's determination that communism in Yugoslavia should be run from Belgrade, not from Moscow, and the reverberations of this defiance spread swiftly throughout eastern Europe. The slightest sign of national independence among communists was condemned as "Titoism," and punished by rigged show trials and death. Since at Stalin's order the Hungarian communist party was playing the part of a loyal member of a coalition bourgeois government, there were ample opportunities for Hungarian comrades to overplay their role, and suffer the consequences. Rákosi himself called the party's strategy "salami tactics," referring to Hungary's favorite delicacy—instead of seizing power in a single revolution, the party would carefully, patiently cut away paper-thin slices of power, a slice here, a slice there, until eventually there was no place in the "coalition" for anybody but communists. The main thing was not to alarm the western powers, or the other parties in the Hungarian coalition government, so comrades who went too far were punished for overzealousness and failure to heed the party line, while comrades who didn't go far enough were punished for Titoism. The penalty for both was the same: torture and death.

The fact that Rákosi and many of his closest collaborators were Jewish of course did nothing to increase the popularity of the Hungarian communist party in a country where, until very recently, anti-Semitism had been official state policy, and by no means an unpopular one until the Germans overdid it. There is some possibility that Stalin, though he was anti-Semitic himself, shrewdly picked Jews to lead the Hungarian communist party precisely because, faced with deep resentment from Hungarians, they would always be obliged to look to Moscow for support, protection, and guidance—there could be no chance that Rákosi and his comrades would "do a Tito" in Budapest. At the same time, the Hungarian communist leaders would feel themselves to be isolated and under siege even when they were victorious and in control of the country, always conscious that hostility toward them, wherever it came from—church, peasants, trade unions, the army—was based on something more than mere politics.

Salami tactics took time—four years, from 1945 to 1949—but they worked. Although the communists never managed to win even 20 percent of the vote, they gradually infiltrated the other parties, and decided more and more of the important issues. By 1947, the AVO (later to become the AVH, but always pronounced by English-speakers as "the Avoh"), Hungary's Russian-trained secret police organization, was busy inventing counterrevolutionary and fascist plots on the part of noncommunist political figures, who were either forced into exile if they were lucky, or lured to Moscow (and from there sent to the gulags) if they were not. A degree of prosperity was ensured by allowing smaller businesses to remain in the hands of their owners, and by not pressing too hard on the peasants to collectivize; and rampant inflation was brought under control, though not without widespread suffering.

Still, by 1947 the mask had been removed. The other political parties were either abolished or incorporated into the Hungarian communist party, a rigged election was held in which the communist party at last—in the absence of any competition or opposition—won over 90 percent of the vote, and Hungary became a "people's democracy" in all but name.

Step by step, the regime increased its pressure on the church (these efforts reached an infamous climax in the trial and imprisonment of Cardinal József Mindszenty in 1949), on owners of small businesses, on landowners and former aristocrats, and above all on the recalcitrant peasants and independent trade unionists.

At the same time, following the model of Stalin, and at his orders, the communist party disciplined and "purified" itself with ruthless efficiency. Those who had joined the party late, and often under compulsion, were driven out of it and purged; there were mass dismissals and arrests of anybody at any level of society who was only a lukewarm supporter of the party line; and a fierce witch hunt was instituted to seek out and destroy communists who might once have expressed sympathy for the Titoist heresy—for it was no longer sufficient to be on the correct side of the party line *now;* it was necessary to have *always* been on the correct side of the party line, even many years ago, as far back as the Spanish Civil War. Fierce denunciations; endless sessions of "self-criticism"; and a detailed and exhaustive study of everything a party member had ever said, written, or thought became commonplace, as the AVO searched, even at the highest level of the party, for the slightest sign of "error."

It is worth noting that Imre Nagy—who would later break from the party, become the leader of the Hungarian Revolution, and seal his own fate by withdrawing Hungary from the Warsaw Pact—was one of those who helped organize the AVO and helped make it a supreme instrument of political repression. Because Marxism was claimed to be "scientific" and because there could therefore be only one scientifically "correct" answer to any question, no shades of gray were permitted, in the present or retroactively, in action or in thought. To deviate, or to have deviated or even to have thought about deviating, from what was now the party line was a crime, all too often punishable by death.

As always in left-wing politics, the revolution devoured its own children. The most important victim of the purges was László Rajk, former minister of the interior and right-hand man

Getty Images

Lazar Brankov and (right) László Rajk, at their trial for treason and espionage, Budapest, 1949.

of Mátyás Rákosi. Rajk was a hardened communist who did not shrink from violence himself; his crime was that he had been a leader of the "underground" communist party in Hungary, rather than being, like Rákosi, a "Moscow man." He had also encouraged a certain amount of populist activity among student groups and peasants, in order to make communism more popular by giving it a more "Hungarian" tone—a tendency which lent itself perfectly to accusations of Titoism.

Rajk was arrested, tried, tortured, and interrogated by his peers (one of them none other than János Kádár, who would shortly be arrested and tortured himself, only to reemerge first as a supporter of Nagy; then as Russia's choice to suppress the Hungarian Revolution; then, after carrying out the mass repressions that followed the revolution, as the artful architect of dismantling communist rigidity). Rajk was executed and buried secretly like many others, a victim of the party to which he had devoted his life, but, in truth, neither a better nor a worse man than those who had disgraced and killed him.

It would be futile to attempt to describe in detail the shifts in party policy in Hungary that marked the last years of Stalin, particularly since much of it was a waste of ink, or blood. The arts, literature, and journalism were severely repressed and controlled, in strict accordance with the Moscow "line," thus stifling what had once been one of the liveliest features of Hungarian life; state-owned industrial concerns replaced private ones; the economy was directed toward heavy industry, leaving the entire country bereft of even the most ordinary consumer items, from clothing to lightbulbs; the hagiography of Stalin was extended to enormous proportions, reaching its peak in the giant statue of the Soviet dictator on Dózsa György Street in Budapest, standing nearly sixty feet high on a rose marble base marked with the inscription, "To the great Stalin from the grateful Hungarian people." Everywhere, the face of Stalin gazed down from framed colored lithographs, usually accompanied by a slightly smaller portrait of Mátyás Rákosi.

In Hungary, as in the rest of eastern Europe, nothing worked. Lines formed twice a day at food shops and bakeries for purchases that were strictly rationed, this in the richest agricultural country in eastern Europe; coal was in short supply, in a country that had always been an exporter of it; trams squealed and rattled through streets and between buildings that still bore the scars and grime of war; cars, mostly Russian Pobyedas, were few and far between, and usually signaled the passage of some official of

AFP/Getty Images

Stalin, reviewing a parade from Lenin's tomb, 1950.

the party or the government, or a sudden, terrifying visit by the secret police. Budapest, which had once been the "eastern Paris," a kind of *ville lumière* on the Danube, was gray, grim, poorly lit; its people were badly dressed, hungry, and living in fear, and with good reason. Informers abounded; the slightest joke about the regime or complaint about living conditions could get you denounced and arrested; and in addition, all those who were of "bourgeois" origin, or had once owned a business or belonged to a "bourgeois" political party, or had served in the Horthy regime, or still clung to their religious belief were subject to harassment, constant suspicion, and the danger of arbitrary arrest.

The numbers are awesome. According to Miklós Molnár, in his authoritative *Concise History of Hungary,* the total of those questioned and charged with a political crime in the years of the repression exceeded 1.6 million, of whom nearly 700,000 were found guilty—this in a country with a population at the time of less than 10 million. Of course Hungary was not Cambodia under Pol Pot—the number of executions was relatively small, and kept rigorously secret—still, all those who were found guilty suffered in some way, as did their families, and the only way to save yourself was to inform on somebody else.

The higher ranks of the party were not excluded. János Kádár, who had helped to interrogate his mentor Rajk,* was himself arrested and tortured—his testicles were crushed—and Imre Nagy, who had helped to found the AVO, was dismissed from the party. Control of the party was tightly held by a trio of sinister

*Miklós Molnár, who is anything but a sensationalist, writes that Rákosi kept a tape recording he had had secretly made of Kádár interrogating his old friend Rajk while Rajk was under torture, as "insurance" against future troubles from the ambitious Kádár. This gives an authentically grisly picture of life in Hungary under the communist regime.

executioners: Rákosi himself, the cold and brutal Ernő Gerő, and the even colder and more sinister Gábor Péter, any one of whom could have taught lessons in cruelty, deceit, betrayal, and hypocrisy to Shakespeare's Richard III. The Hungarian communist party was in some respects more Stalinist than Stalin himself. Certainly it obeyed Stalin slavishly, unlike the Polish party, which, while it did its best to obey the Kremlin's wishes, was nonetheless deeply ambivalent about trying to turn Poland—a fiercely anti-Russian country, deeply religious, with strong trade unions—into a replica of the Soviet Union. Rákosi had no such doubts, and would allow none in the party—or indeed in the country.

Getty Images

Nagy addresses Parliament. Glaring at him on his right, Rákosi; on his left, with a sinister, sardonic smile, Gerő.

Dissent was stifled by terror, the arts were targeted by the party to suppress anything but the most slavish imitations of Soviet socialist realism, and the media were systematically brought under strict party control. Even the smallest of private businesses—cobblers, tailors, lawyers, and doctors—were abolished, so that professional people and small businessmen joined the former aristocrats in labor camps, or doing manual labor, with the predictable result that nothing could be done that required any degree of skill or knowledge, from repairing a clock to putting new heels on your shoes.

This period ended abruptly with the death of Stalin, in March 1953. Events in the Soviet Union moved rapidly though, for those whose lives depended on guessing the outcome of the inevitable struggle for succession, with much confusion. Lavrenty Beria, whose power, as head of the secret police, was second only to Stalin's, was unexpectedly denounced by his colleagues, arrested, and executed. A "collective" leadership was proclaimed, split between the older figures around Stalin, men like Molotov and Mikoyan, and their younger rivals, party apparatchiks and yes-men to Stalin, like Georgy Malenkov, Nikita Khrushchev, and Nikolay Bulganin. Rumors spread that the power of the secret police would be curbed, that "socialist legality" would replace outright terror, and that Stalin's "excesses" would be ended; but all of this was hard to decipher, even for the most astute of Kremlin watchers, since it all took place behind a somewhat unconvincing display of solidarity among Stalin's heirs. There was talk of a thaw in the Soviet Union, though whether this would extend to the people's democracies of eastern Europe was unclear.

Then, in June 1953, the Hungarian leaders were called to Moscow. Experienced Stalinists, they must have approached the journey with deep misgivings, and wondered who among them would fail to return. To everybody's surprise, Mátyás Rákosi was removed from the government, though left in control of the party; his place was taken by Imre Nagy, whom Rákosi had disgraced and driven out. It did not escape people's attention in Hungary that fearsome as the Hungarian communist leaders were, they could be summoned to Moscow and moved around like so many chess pieces by the Russians. That the Russians' choice fell on Nagy came as a relief to everybody but Rákosi and his henchmen. Nagy was a "moderate"; he had opposed forced collectivization of agriculture; and he had the look of a man who enjoyed his food and his wine, rather than the ascetic, grand inquisitor look of, say, Gábor Péter.

Nor was Nagy's plumpness and bland smile a disguise—he *was* a bon vivant; he enjoyed dancing with pretty girls in peasant costume and was photographed doing so; his ambition was said to be to put a "human face" on communism, and he proceeded to do that with amazing speed. Attempts to gather the peasants in collective farms or, still worse, *zovkhozes* on the Soviet model, which were as rigidly organized as the army, were slowed down; plans for even bigger industrial collectives were shelved; people detained by the secret police for political crimes were released; investment in consumer industry was increased. It was not quite yet "goulash communism," but it was a start, and overnight Nagy became very popular indeed. And indeed there was something attractive about him, stiff and awkward as he sometimes looked, a neatly tailored figure, leaning back with a stout belly thrust forward, rather like a combination of a penguin and the Michelin tire man. He spoke like a professor, but his speeches were laced with earthy peasant phrases, and while he was no spellbinding orator, he always managed to sound sincere. He was a decent man, reputed to have a good sense of humor, unusual character traits among the scowling brutes who had hitherto been running Hungary on Moscow's orders, and who now hated Nagy and waited impotently for their revenge.

Nagy ran Hungary for almost two years, and "lifted the lid off the pot" just enough to let a little of the steam out. Life was better, and food more plentiful; though Hungary was still a police state, the worst excesses of the terror were ended, and a certain freedom of thought in the arts was permitted once again. Best of all, Nagy seemed to have the gift of doing all this without alarming the Kremlin. Georgy Malenkov, the plump apparatchik whom Stalin had so often terrified and humiliated at the dinner table with coarse and threatening humor, had emerged as the Soviet leader, and he and Nagy seemed made for each other, both fat men who enjoyed a good meal, and above all wanted to combine communism with a certain degree of bourgeois comfort.

This comparatively benign period ended when Nikita Khrushchev succeeded in ousting Georgy Malenkov. Khrushchev, despite his gold-toothed smile, his portly physique, and his red cheeks, was an altogether tougher person than Malenkov. Khrushchev had "made his bones" (as they say in the Mafia) in the communist party by his energy in forcing the Ukrainian peasantry at gunpoint into collective farms and executing in huge numbers the kulaks, which is to say the wealthier and more successful farmers. It was his success at both these enormously cruel tasks that first brought the young Khrushchev to Stalin's attention. Khrushchev had mastered to a high degree the art of looking jolly,* but a close look at his eyes told a very

*As a Russian-speaker, I was one of the undergraduates chosen to translate for Khrushchev and Bulganin when they visited Oxford in 1954, and therefore had a good opportunity to observe Khrushchev close up. When he came to Magdalen College—my own college—he was taken to see Magdalen Tower, one of Oxford's most admired pieces of architecture. The president of the college, T. S. R. Boase, explained to Khrushchev that it was an Oxford tradition, hundreds of years old, that on the first of May every year, the Magdalen

different story, though of course this is not to say he was another Stalin.[2]

On the other hand, eager as he was to exorcise the ghost of Stalin, he was a shrewd judge of power politics; he recognized the Western recognition of the Federal Republic of Germany in the autumn of 1955, and the swift rearmament of Germany, as a signal to rein in the Soviet satellite countries before they got out of hand. Much as the eastern European countries might wish otherwise, they were needed to help defend the Soviet motherland.

Khrushchev had his own problems with the Soviet generals, not all of whom were eager to bury Stalinism along with Stalin, and at the very least he did not want to be blamed by them for any lukewarm response to Soviet demands for more troops. The formation of the Warsaw Pact was the Soviet reply to an independent West Germany, and the people's democracies were brusquely ordered to expand their armies and invest in the heavy industry that was required to equip them with modern weapons. Under the circumstances, Nagy's reforms in Hungary were suddenly seen to have been a movement in the wrong direction, and he was swiftly removed from office and replaced by his old nemesis Rákosi.

Nagy was lucky that his life was spared—if the decision had been up to Rákosi and Gábor Péter, he would no doubt have been executed; but in Moscow it was thought sensible to keep him alive in case Rákosi made difficulties. In the meantime, Hungary

---

School choir sang hymns from the top of the tower at dawn, and that thousands of people waited all night, in a festive atmosphere, to hear them. Khrushchev stared up at the tower belligerently. "Why?" he asked, with great impatience. Then, turning to Bulganin, he said irascibly, "Can you believe how fucking superstitious these people are?" Tactfully, I did not translate this remark for the benefit of President Boase.

was back in the deep freeze of the cold war, and would remain there, a victim of the changed balance of power between East and West, because an armed, prosperous, and democratic West Germany was a direct threat to the Soviet Union's hegemony in eastern Europe—one only had to compare East and West Berlin at the time to see the contrast.

A Visit behind the Wall

I no longer remember what inspired Michael Maude and me to drive through eastern Europe in his new MG. Maude, like myself, was partly American (his mother was an American), without being especially Americanized—indeed, he had done his national service in the British army as a subaltern in one of the regiments of the Foot Guards (I forget which), in part because it had been the old regiment of his grandfather, a general in World War I.* Like me, Michael Maude had served in the British zone of West Germany, but had never been across the border into East Germany, where we were absolutely forbidden to go, and this may have been the reason for our trip behind the Iron Curtain, or perhaps he only wanted to show off the MG to the East Germans.

My father was generally in favor of any kind of travel scheme, however far-fetched, but he was uncharacteristically unenthusi-

*General Maude was famous in the British army, since he had lingered too long on the beach, which had been mined, looking for his luggage, during the evacuation of British and ANZAC forces from Gallipoli, prompting a literary-minded officer to recite, in a parody of the well-known poem by Tennyson:

> *"Come into the lighter, Maude,*
> *For the fuze has long been lit,*
> *Come into the lighter, Maude,*
> *And never mind your kit."*

astic on the subject of our visiting Berlin, Dresden, Leipzig, and Prague. He wondered aloud whether a trip through the wine country of France might not be more amusing and safer; eastern Europe, he thought, would be "very sad." But eventually he gave us his rather lukewarm blessing, and made a few calls to friends such as Churchill's old pal and financial adviser Brendan Bracken, to make sure that the Foreign Office was alerted about our trip, in case of trouble. Graham Greene, another family friend, also consulted by my father, thought it sounded like grand fun, and I think if the MG had had more than two seats he might have come along.

Maude had come into possession of all the "kit" his grandfather had been searching for on the mined beach at Gallipoli—it filled his sitting room overlooking Deer Park at Magdalen, and we could see why the general had been unwilling to leave without it. Here was everything an Edwardian officer might ever need for a campaign anywhere in the world, all of it bound in gleaming leather and green waterproof canvas, and marked with General Maude's initials: trunks, shotguns, binoculars, folding bed, portable toilet, Sam Browne belt, sword and scabbard, a Mauser "broomhandle" pistol, canteens, a campaign kitchen set with monogrammed china and silverware, hatboxes, flasks of every type and size, a vast pile of luggage that must have caused many a soldier or coolie to sweat and swear and many a mule to protest while carrying it. Maude wisely left the weaponry behind, but even so, once his luggage had been strapped to the boot of the MG and covered with a rainproof tarpaulin, it seemed unlikely that the car would be able to climb even the smallest hill.

In those days, a Channel crossing was still something of a maritime adventure, on ferries that had once upon a time served to evacuate the BEF from Dunkirk. We drove quickly from Calais to the Belgian border—this was in the days before the grand autoroutes were built, when you still drove through France and Belgium on slick two-lane roads and skidded and bumped through every town over cobblestones and tram rails—surrounded by prosperity, crossed Belgium and Holland, then crossed into West Germany without drama, and zoomed onto the autobahn. West Germany was awash with prosperity; everywhere you looked there were new Mercedes and Volkswagens galore, until we reached the Iron Curtain itself at Marienborn.

Here, a whole different atmosphere awaited us. The East German border guards and the scowling Volkspolizei looked as if they were wearing old SS uniforms with the insignia removed, and their attitude was frosty, unfriendly, and suspicious. You waited, in a room smelling of damp wool, boot polish, and sausages, while every page of your passport was scrutinized. All Maude's luggage had to be carried inside, unstrapped, opened, inspected. It did not help matters that we were keen amateur photographers—the serial number of every camera and lens had to be written down meticulously on the back page of our passports and on a list kept by the police.

I realize now, of course, that our own appearance did us no good in their eyes. Maude was wearing his officer's "British Warm" overcoat, and I was wearing my old R.A.F. leather flying jacket, with a fleece lining and collar. With our polished shoes, tweed sports jackets, narrow cavalry twill trousers, Viyella shirts, and striped ties, we could not have looked more like two young British military men in civilian clothes, setting off for a clandestine photographic journey through the German Democratic Republic. We might as well have been in full uniform. Even the

MG inspired suspicion, since it resembled a sporty stage prop, with its chrome wire wheels and fold-down windshield.

In contrast to those in West Germany, the East German autobahn was empty, except for the occasional Soviet military convoy. The MG seemed to cheer the Russian soldiers up no end—perhaps they thought that if this was the best Western technology could produce in the way of a car, the cold war was as good as won. We spent a night in West Berlin—which was then rather like Broadway and Forty-second Street with a German accent—and crossed into East Berlin the next morning.

This, we quickly realized, was the real thing: gray, dingy, dreary, still full of huge piles of bomb rubble, the horizon dominated by hideous new gray cement buildings and giant, hideously ugly memorials to the Soviet "liberators." The wide avenues were empty of traffic; the people were shabby and unwilling to look in our direction; the police presence was overwhelming; the shop windows were empty. It was the world George Orwell had described in *1984*. But Dresden, when we got to it, made East Berlin look like Paris. Thanks to the Royal Air Force and Air Chief Marshal "Bomber" Harris, the head of Bomber Command, 1,000 bombers had leveled it in a firestorm, in the spring of 1945, leaving behind a charred and blackened moonscape. Here we were treated not only with suspicion but with outright hostility, as representatives of the country that had bombed and destroyed one of Europe's most beautiful cities. Cold war or no cold war, here the British were still the enemy. It was only with great difficulty that we were able to find a restaurant that would serve us at all—even though the restaurants were empty, all the others had claimed to be "booked up"—a huge, institution-like place, with ghastly food and weak beer. In Leipzig, we were almost arrested by the Volkspolizei for taking a photograph of the MG outside the railway station, and all the way to the Czech border we were followed by a succession of shabby cars full of plainclothes policemen.

Prague, when we got to it, came as a relief. People were at least pleased to see two Englishmen with pockets full of hard currency, and at last there was something to buy, if you liked glassware. Alone of central European cities, Prague had survived the war unscathed, and still showed signs of its former beauty, but it is almost impossible to describe the drabness, the poverty, the lack of neon or bright lights, the emptiness of the streets, and the constant ominous presence of the police. Even the purchase of a postage stamp caused suspicion and fear on the part of the postal clerk, as if selling a Czech stamp to a foreigner was a crime.

To arrive back in the West—in Munich—was like returning from Mars. We had contemplated going on to Budapest, but in Prague we decided to skip it. As usual, my father had been right—eastern Europe was a very sad place to visit. Behind the Iron Curtain was where so much of the twentieth century had gone hideously wrong. The scars of war were still unhealed; the miasma of guilt, betrayal, and anger was as unmistakable as the gritty, coal-specked mist and the smell of old brick dust and fire. Long before John le Carré wrote *The Spy Who Came in from the Cold,* it was the atmosphere that he would capture so perfectly, and that would permeate so much of his best work.

Eastern Europe was stuck—stuck in the past, stuck in 1945. Although the Russians would soon be testing their own hydrogen bomb and putting the first satellite into orbit, there was nothing they could do to improve the lives of the people who lived in the countries they controlled, for they too were stuck—stuck with a rigid ideology that admitted no deviation or error;

stuck with a doctrinaire view of economics that would eventually lead them nowhere; stuck with Marxist, state-run plans for growth that would eventually sink the Soviet Union and all the people's democracies beneath the weight of their own heavy industry, while remaining unable to satisfy the consumer needs of the population. The party's need to maintain a police state; its stranglehold on artistic expression and opinion; the ever-increasing demand for obedience, discipline, and orthodoxy—all of these things made it impossible for them to compete with the West. Marx, Lenin, and Stalin had led them into a blind alley.

Hungary was no exception.

In April 1955, Nagy was removed from office, and all power was once again in Rákosi's hands. Nagy was even expelled from the communist party, and obliged to hand over his party card—the ultimate humiliation. Yet here, too, there was a sign of weakness. Stalin would have had Nagy executed; but instead Nagy merely retired to his country house to live comfortably.

Rákosi recklessly plunged Hungary back into a political deep-freeze, even as the Soviet leadership performed a ritual journey to Canossa by flying to Belgrade to apologize to Tito. Thus Soviet policy, perhaps inadvertently, undercut the position of communist leaders in the satellite nations. This very dangerous fault line widened abruptly in February 1956, when Khrushchev made his famous speech at the Twentieth Party Congress denouncing Stalin and Stalin's crimes. In Poland, Hungary, and Czechoslovakia Stalin was still being praised and Stalinism was alive and well; in Moscow Stalin was being reviled. Future criticism of Khrushchev for being "impulsive" scarcely does justice to the imprudence of this line.

Getty Images

Stalin's heirs: the Soviet Politburo and distinguished visitors, on Lenin's tomb, Moscow, celebrate the anniversary of the October Revolution.

The consequences were quick to arrive. In Poland the trade unionists made demands that the government would not meet; they went out into the streets of Poznan to protest, and were promptly shot down by the "security forces." In Hungary, the "Petőfi circle" of anti-Rákosi communists and intellectuals held debates on the most controversial of social issues, some of which were broadcast into the street by loudspeaker. Rákosi wanted the Petőfi circle crushed, but the Soviet leadership, appalled by what seemed to be the first act of a proletarian revolution against the communist party in Poland, instead sent Anastas Mikoyan to Budapest, first to calm down Rákosi; then, when this proved impossible, to replace him with Ernő Gerő, another murderous brute, but at least a different and somewhat chastened one.[3] Yuri Andropov, future head of the KGB, who was then the Soviet Union's ambassador to Hungary, informed Gerő that Comrade Rákosi would be kept in the Soviet Union for as long as necessary, and advised helpfully that the Hungarian press should announce he was receiving "prolonged medical treatment."[4]

On the other hand, the damage had been done. The Russians had urged conciliation on the shaken Polish government; Mikoyan had removed Rákosi; the lid was off the pot.

It was too late. The contents were about to boil over.

5.

"Arise, Magyars!"

As everybody who reads history knows, it is precisely the moment when those in power seek to reform or ameliorate conditions that leads to revolution—a fact that Louis XVI ignored when he summoned the Estates-General to Versailles to discuss tax reform in 1789. It should have come as no surprise, therefore, that the Soviet Union's attempt to smooth matters over in Poland—by forcing the reluctant Polish government and party to accept the dissident Wladyslaw Gomulka back into their ranks—and Anastas Mikoyan's dismissal of Rákosi in Hungary had an effect that was diametrically opposite to what was intended. Far from calming things down, it heated things up.

In Hungary, the immediate result was a demand for a state funeral for László Rajk. The government gave in to this demand, and the funeral suddenly blossomed into a huge, spontaneous public event, unprecedented in a communist country. Mrs. Rajk—who behaved with great dignity—was accompanied to the cemetery by a procession of over 100,000 people, including many of those who had helped condemn Rajk, as well as many of Rajk's victims, marching silently through the streets of Budapest, while the AVH and the police looked on impotently.

This was followed very shortly by an announcement of the demands of the intellectual Petőfi circle and various student groups—including the withdrawal from Hungary of Soviet troops, the cancellation of further reparations payments to the Soviet Union, the return to power of Imre Nagy, and free elections.

On October 23 an immense procession calling for the implementation of these twelve or sixteen demands (the number varied according to the speaker) set out to honor the heroes (and martyrs) of the revolution of 1849, Petőfi and General Bem. Imre Sinkovics, an actor from the Hungarian National Theater, recited the words of Petőfi's "National Poem" beneath his statue (they had lost none of their ability to arouse patriotic fervor to a white-hot intensity over the last 108 years), and the crowd, now over 300,000 strong, began to move through the streets of Budapest.[1]

The members of the government briefly watched this astonishing scene, discussed using the security police to open fire on the demonstration if it got out of hand, but then came to the alarming conclusion that an order to fire might not be obeyed. Once again, the fear of using force doomed those in power—the communist party and government in Hungary in October 1956 were as helpless as the monarchy of France had been in 1789 or the czar in 1917. The crowd began to tear the Soviet symbol out of the center of the Hungarian flags they were carrying; this was in accordance with demand number fourteen of the Petőfi circle's manifesto: "the replacement of emblems that are foreign to the Hungarian people by the old Hungarian arms of Kossuth." They then began tearing down communist stars, emblems, and flags from the buildings they passed, with the help of the firemen of Budapest, who brought their trucks and ladders to reach the ones that were high up.

Growing in numbers, one part of the crowd moved to try to seize the radio station, while others marched on the newspaper offices and still others to the giant statue of Stalin. The last group, with immense effort (and the help of a passing truck carrying a welder and his tools), finally succeeded in toppling the statue, leaving behind on top of the rose marble plinth only two giant bronze boots over six feet high. There was very little disorder, despite later claims to the contrary, and no looting. Such disorder as there was consisted of people bringing framed and unframed portraits of Lenin, Stalin, and Rákosi out into the streets and burning them, along with these men's collected writings. The occasional bonfires had a certain sinister quality, however, as if presaging the shape of things to come.

The situation was now completely out of hand. The statue of Stalin was dismembered; people stopped in the street to spit on its giant head; stonemasons appeared from nowhere with their

Popperfoto / ClassicStock

Budapest, October 1956: the head of Stalin's giant statue in Budapest lies in the street.

tools to obliterate the carved inscription on the plinth of the statue; the AVH and the security forces either attempted to vanish into civilian clothes or barricaded themselves in their headquarters to defend their lives. In their absence, the authority of the government was, for all practical purposes, at an end. As in a city in the midst of a general strike, all services ground to a halt. The crowds increasingly took up the chant *Ruszki, haza! Ruszki, haza!* ("Russians go home!"), and began to call for Imre Nagy, who had returned to the city from his home on Lake Balaton.

Although the hard-liners of the central committee—men like Gábor Péter and Ernő Gerő—were, now that it was too late, anxious to use force to "restore order," it was increasingly obvious that this could be done only by calling in the Russians. By the middle of the afternoon, military cadets, policemen, and soldiers were already marching with the crowds of demonstrators and tearing the Soviet stars and emblems off their caps and uniforms. Already, weapons were appearing in the hands of civilians. The party had stockpiled plenty of weapons around the city in factories and government buildings, to use in case of a challenge to the party's rule, and these were now about to be used against it—in a country where military service had been compulsory for years, there were very few people unfamiliar with the use of weapons.

By the early evening a mass of people—perhaps as many as 250,000—had assembled in Parliament Square, shouting for Nagy to appear. At first the government attempted to disperse the crowd by turning the lights off in the square, but the demonstrations grew so unruly that they were switched back on again. Eventually, sweating and nervous despite the cold night air, Nagy, who had been persuaded with great difficulty to address the crowd, appeared on a balcony, and was cheered for many minutes. "Comrades . . . ," he started off, but the people shouted him down with cries of "No more 'comrades'!" Nagy, who had been a communist all his adult life, had to start again, with some embarrassment, addressing them instead as "young Hungarians."

In the meantime, a separate drama was taking place outside the government radio station, where a recorded speech by Ernő Gerő was about to be played. Gerő's speech was harsh, threatening, and unconciliatory, and it sparked off the first moment of real violence. The crowd demanded a microphone to broadcast their demands and repudiate Gerő's speech; the woman director of the radio station refused; people began to throw bricks through the windows of the building; and then somebody backed a van into the big wooden gates at the entrance and knocked them down. At this the AVH guards in the building fired tear gas into the crowd and opened fire with automatic weapons.

The party and the government were aware, from the very beginning, of the profound importance of the radio station. A modern revolution requires control of the media, and a radio station is a more significant weapon than guns. The building had been garrisoned with AVH guards, and in the streets outside were tanks and troops of the Hungarian army. Perhaps the most critical moment of the revolution was now about to occur. As the hated AVH fired into the crowd, the soldiers in the trucks or standing on tanks began to hand their weapons down into the crowd. The weapons were passed overhead from hand to hand, gleaming dully in the light of the streetlamps, until they reached the people in front, who began to fire back on the AVH. The crowd was packed tight. There was no way to duck or lie flat. People squeezed and pushed each other to make way as the wounded were dragged through to the outer edges of the crowd. Medical students in white coats appeared to help them. At last those of the crowd who were armed managed to break into the radio building. AVH members who were trying to change into stolen civilian clothes were shot, as were some who had hidden in the women's lavatory. Others were thrown out the windows into the street, and killed there.

Similar scenes were taking place in the newspaper offices and elsewhere, although the headquarters of the communist party was too strongly defended by tanks for the crowd to reach.

At the party headquarters, the senior members of the party and the government, many of them by now scared out of their wits, struggled to decide what to do. Gerő's speech was widely criticized, even by hard-liners, as having been injudicious and inflammatory, but with the radio station now in the hands of the insurgents it no longer mattered. In the end two decisions were made that were mutually contradictory and reflected a certain fatal ambivalence: to call Moscow and ask for Russian troops to restore order, and to appoint Imre Nagy prime minister.

The call to Moscow, if it was ever made, was unnecessary—Moscow had already made up its mind. At a meeting of the Presidium of the Central Committee of the Soviet communist party, Nikita Khrushchev had already spoken out in favor of military intervention; Russian troops were already entering Budapest and were moving to secure every important point in the country. As for Nagy—who was hastily reinstated as a party member and a member of the Hungarian Central Committee—he would deny to the very end that he had been involved in calling the Russians in. His immediate concern was to put together a government that contained enough "real" communists to satisfy the Russians and enough "reform" or "liberal" communists to pacify the rebels in the streets.

But it was already too late. As dawn came, fighting was widespread. Soviet military aircraft flew overhead, and Soviet tanks and armored personnel carriers moved through the city in large numbers—it was an impressive show of force, but it was not sufficient to produce a knockout blow. The Soviet forces in Hungary were large, but they had a whole country to occupy, and for the most part they consisted of conscripted soldiers—the Soviet equivalent of Britain's "national servicemen"—armed with weapons left over from World War II. There were plenty of tanks, but they were mostly T-34s, not the newer T-55s, along with some very much heavier JSIIIs—then the world's biggest, heaviest tank, really too big for street fighting—and a mixed bag of self-propelled heavy artillery.

Still, given their numbers and their preponderance of heavy equipment, the Soviets ought to have been able to put down the uprising; but they badly underestimated the anger of the insurgency and its determination to fight, as well as its advantage in knowing every alley, street, and building in Budapest.* The fact that so many of the fighters were young also meant that they were willing to take risks that might have seemed unimaginable to an adult. The techniques the young people—many only schoolchildren—used to destroy Soviet tanks were amazing. The most reliable and widespread was the famous Molotov cock-

*It is difficult to estimate how many Hungarians took part in the fighting, since many fought, then gave up their weapons to somebody else, then rejoined the fighting later. A figure of 4,000 in Budapest (and perhaps as many again in other major cities) is often given, but that probably represents only the number of fighters actually armed and in the streets at any one time. The total number of people who took part in the fighting at one time or another during the two phases of the struggle would be far higher. C. Philip Skardon, an analyst at the CIA at the time, put the total number of fighters at 15,000, of whom 2,502 were killed and almost 10,000 wounded, and this seems plausible. The number of those executed afterward was in the low hundreds.

tail—an improvised antitank weapon, originally devised by the Soviets, consisting of a liquor bottle full of gasoline with a gas-soaked rag stuffed into the neck of the bottle. To use a Molotov cocktail, it was necessary to sneak up close to the tank, run toward it from behind (largely a blind spot for the tank crew), light the gas-soaked rag, then fling the bottle upward from a distance of less than six feet so that it landed on the cooling vents of the engine compartment, dousing the engine with blazing gasoline. Either the engine caught fire, forcing the crew to abandon the tank by climbing out of the hatches (and be shot down by older insurgents from nearby doorways), or the tank blew up ("brewed," as we used to say in the British armed services), incinerating the crew.

There was never a shortage of empty liquor bottles, of course, or, as it turned out, of young people to throw them at tanks. As for the tank crews, they hated city streets, particularly the narrower ones, where every doorway and window might hide a kid with a Molotov cocktail, and where everything from felled streetlamps to trucks and streetcars could be used to create improvised barricades that would slow down a file of tanks or stop it altogether. Tank crews like open country, with areas of good cover from which to shoot; and like all soldiers they have a particular dislike of fighting against amateurs who don't take prisoners. And from the very beginning, neither side was inclined to show mercy. Anybody who tried to surrender was likely to be shot, and while the medical students very bravely picked up the Hungarian wounded under fire, they did not bother trying to save the Russians, who, for their part, didn't pay much respect to people in white coats, or to ambulances marked with a red cross.

From dawn on day one, the Russians used their artillery at the least sign of resistance. Since the Russian army in Hungary had never been intended to fight NATO—it was primarily an army of occupation—the tanks carried high-explosive ammunition for use against buildings and fortifications, rather than armor-piercing rounds for fighting against other tanks, and this did enormous damage to buildings.* Budapest had been remorselessly shelled during the Russian siege in 1945, and now it was to be shelled again.

If the Kremlin had anticipated that a demonstration of brute force would calm things, it was badly mistaken. The effect of Soviet tanks in the streets of Budapest (and other cities) was like a bear smashing its paw into a wasps' nest—even sincere communists were likely to pick up a weapon and fire back. The Hungarians were divided about many things, but they were determined to resist the Russians with lethal violence. To the horror of the Kremlin, Imre Nagy, and the Western powers, the morning of October 24 saw Budapest in the throes of a full-scale battle, in the midst of which the soothing words on the radio announcing that Nagy had been appointed chairman of the Council of Ministers—in effect, prime minister—had very little effect.

Except for the appointment of Nagy himself, the new government pleased nobody, since Ernő Gerő, a hated and reviled Stalinist, was first secretary of the party; and the hard-liner András Hegedűs was first deputy chairman of the government. In short, Nagy seemed to be hemmed in between two of Rákosi's

*Later, when the Russians came back into Budapest with first-rate, well-equipped armored divisions from the Ukraine, they left their bases so quickly that they had no time to exchange the armor-piercing (AP) shells in their tanks for high-explosive ones. These AP rounds tended to make a neat hole in a building before exploding inside, rather than bringing the whole side of a building down into the street, and were nothing like as effective against makeshift barricades.

closest collaborators. In the streets, shouts and placards with the slogan "Death to Gerő!" began to alternate with "Russians go home!"

The Russians' plan had been to seize the key points in the city quickly—the bridges, the main street junctions, the telephone exchange, the barracks, the railway station—but by midmorning they were bogged down in fierce firefights throughout the city, for the insurgents were being joined by the Hungarian army, with its own tanks and antitank guns. At the same time, a general strike was spreading, paralyzing the country as trains came to a halt, and adding large numbers of armed workers to the ranks of what the world press was already calling the Freedom Fighters.

At noon, Nagy broadcast an appeal for a cease-fire, promising that all those who laid down their arms before two in the afternoon would be spared. He urged a quick "return to peaceful and creative work," but by then the fighting in the streets was out of control, with tank battles taking place all over the city. Nobody laid down arms at two o'clock—instead, a rather more sinister event took place: the arrival at party headquarters of Anastas Mikoyan and Mikhail Suslov, both Soviet deputy premiers, who had flown from Moscow and been brought from the airport in a heavily escorted Soviet armored vehicle. Mikoyan was wily, tough, and shrewd, one of Stalin's most resourceful trouble shooters, Suslov was the party theoretician and a confirmed hard-line Stalinist. Both men were appalled by what they saw, and by the paralysis of the Hungarian government and party, which they blamed on Gerő rather than Nagy.

Fighting raged through the day and on into the night, while the government tried and failed to persuade the Hungarian army generals to break off their support of the insurgents. In the evening János Kádár—who still had, in most people's eyes, a certain prestige from having been tortured at the height of Rákosi's purge of the party, and as a result had been made a member of the Hungarian Politburo and a secretary of the Central Committee—made a broadcast, calling the Russians "our brothers and allies," which had a very poor effect. Gerő, Nagy, and Kádár had all three made broadcasts in the past twenty-four hours—Gerő's was widely held to have exacerbated the insurgency, and certainly neither Nagy's nor Kádár's had calmed it down.

Time Life Pictures/Getty Images

Kádár.

John Sadovy / Time Life Pictures / Getty Images

Colonel Pal Maleter, on the right, wearing a tanker's helmet.

There was now a curfew in effect, as well as a general strike. At dawn on October 25, people came out into the streets again, to a city that already had the appearance of a battlefield. The first news of the day was that Colonel Pal Maleter, a six-foot-six Soviet-trained tank officer, had taken command of the Kilian barracks and gone over to the insurgents. The next was that Nagy would speak to the people from Parliament—Nagy's prestige, despite his being part of the government that had called the Russians in, was such that thousands of Hungarians began to march to Parliament Square to hear him. Russian tanks, alarmed by the crowd, moved to block the way to Parliament. As those in the crowd who knew Russian tried to harangue the tank crews, AVH soldiers stationed on the roof of the Parliament building opened fire into the crowd. Startled, the Russian tank crews opened fire too, and the crowd—which numbered between 30,000 and 50,000—was caught in a vicious cross fire, which killed between 300 and 500 men, women, and children. This set off a day of confusion, recrimination, and further bloodshed. Gerő was expelled—too late—from the government and the party; Kádár and Nagy both made conciliatory speeches (in later years it would be alleged that Nagy was forced to rewrite his speech at gunpoint, on the orders of Mikoyan and Suslov); a crowd of demonstrators gathered outside the British legation to beg for help from the West and received the somewhat tepid condolences of the British minister; and President Eisenhower, nearing the end of his campaign for a second term, expressed his sympathy for the Hungarians, who might have been disappointed had they known that in Washington the Special Committee on Soviet Problems had debated whether the president should announce a special day of prayer for them.[2]

Sympathy and perhaps prayers, it was already beginning to appear, were all they were likely to get. For years Radio Free Europe, broadcasting from Munich, had been urging the people of the "captive nations" to rise against the Russians, and promising help from the West when they did so. Much as these broadcasts irritated communist governments, it is doubtful that anyone else took them all that seriously. America had not come to the aid of the martyred strikers at Poznan in Poland, nor did anybody expect it to. Now, despite the fulminations of Secretary of State John Foster Dulles and the right-wing Republicans against the Soviet Union, Eisenhower tried to make it as clear as possible that anyone who was waiting for the appearance of

American tanks on the Danube to support the insurgents would be waiting a very long time indeed.

At the same time, events in the Middle East were conspiring to prevent any meaningful attempt to restrain the Soviet Union. The British and the French, acting in concert for once, but with extraordinary (and ill-fated) duplicity, had schemed with the Israelis to encourage a surprise attack by Israel on Egypt. This attack would serve as the pretext for an Anglo-French invasion to "stabilize" the situation in the interests of world peace. Behind this fig leaf was an old-fashioned nineteenth-century imperialist plot: the English and French planned to remove President Nasser of Egypt from power; return the Suez Canal, which he had nationalized, to Anglo-French ownership; and put a pro-Western government in power in Cairo.

Alas, neither Anthony Eden nor Guy Mollet was an imperial adventurer in the tradition of Kipling's heroes; nor were France and Great Britain great powers anymore. President Eisenhower, infuriated at having been duped by America's closest allies, refused to countenance the Anglo-French invasion of Egypt, and moved to punish the French and the British where it hurt, cutting off dollar credits and oil supplies. Egypt would be bombed, and British and French troops would go in; but at the insistence of the United States these troops were quickly withdrawn, and any claim Britain and France might still retain to the position of great powers vanished, along with the careers of the two statesmen.

That this drama—or farce—was being played out at the same time as the Hungarian uprising was a tragedy, first of all because it gave the Soviet Union a moral fig leaf of its own—why should it be condemned by the West for doing just what the French and the British were doing in Egypt?—and second because it diverted attention from Budapest to Egypt. Everybody feared that the Soviet Union might airlift massive numbers of troops to Egypt to support President Nasser, or even use nuclear weapons.

By now it has become second nature on everybody's part to deny that American and British intelligence had anything to do with events in Hungary; but fifty years later there is good reason to believe that this denial is false. The British had every reason to encourage a rising behind the Iron Curtain as a way of keeping the Soviet Union and its army tied down in the streets of Warsaw or Budapest, as opposed to rushing to the support of Nasser. One CIA analyst at the time, C. Philip Skardon, in an excellent monograph on the subject, *A Lesson for Our Times: How America Kept the Peace in the Hungary-Suez Crisis of 1956*, traces in some detail the behind-the-scenes intelligence activity that preceded the Hungarian Revolution, and concludes from intercepts that British intelligence (MI6) was deeply implicated in encouraging student groups and intellectuals to revolt, and in persuading anti-Rákosi members of the party that a revolt would receive backing from the West. The French and the Israelis left this side of things to the British, who had always had good contacts in Hungary, the idea being that the two crises would occur at the same time, which was exactly what happened.

Apart from Radio Free Europe, the American intelligence community was deeply resentful of British attempts to influence events in Hungary, and appalled by the prospect that the United States might have to intervene in two armed conflicts, both in areas where the Soviets' sensitivity was acute and where American involvement might easily provoke the Kremlin into the use of nuclear weapons, with unforeseeable results. Something of this attitude was reflected in Eisenhower's actions—he simultaneously halted the Israelis, the British, and the French in their tracks and made it clear to the Soviets that the United States would not cross the Iron Curtain to support the Hungarian insurgents, thus defusing the most dangerous moment of the cold war until the Cuban missile crisis.

There is ample reason to believe that without the Suez crisis, the Hungarian revolt might have had a happier ending. America could have put greater pressure on Russia than it did, the British and the French could have joined in usefully, and some kind of negotiated international settlement might have been reached; but the fighting in Suez precluded all that, and the Soviet Union was left free to devote its full military strength to ending the Hungarians' bid for independence.

Sipa Press

Lynched AVH officer, Budapest.

. . .

None of this was obvious on October 26, and the Budapest correspondent of Britain's communist newspaper *The Daily Worker* was accordingly inspired to write a long, truthful account of what was taking place in Hungary, which, naturally, *The Daily Worker* spiked. Fighting had by now spread throughout the country. In Győr and Magyaróvár, near the Austrian border, frightful massacres took place, carried out by the Russians and the AVH, followed inevitably by the ruthless murder of all AVH troops and agents, many of whom were lynched. In places, the bodies of AVH men were stripped and hung up from trees; in Budapest the photographer John Sadovy of *Life* took a famous series of pictures of the execution in the street of captured AVH guards. All over the country, hastily improvised funerals were taking place, of men, women, and children killed during the fighting. In the streets, Russian corpses were separated from those of Hungarians. Medical workers sprinkled the bodies with quicklime to prevent infection, but the corpses of Hungarians usually bore a small bouquet of flowers on the chest, and sometimes their names, hand-lettered on a piece of cardboard. The bodies of AVH informers and troops often had a framed portrait of Stalin or Rákosi on their chest, to set them apart from Freedom Fighters and civilian victims of Soviet aggression.

The government's ability to handle the situation was visibly deteriorating rapidly—food supplies were erratic, with people queuing up for bread while fighting was going on around them; hospitals were filled to overflowing and running short of all kinds of medications and supplies. The Budapest radio station was back in the hands of the government and the Russians, but many other radio stations across the country were in the hands of the insurgents, so radio news programs broadcast two vehemently opposed views of events.

October 27 was another day of violent fighting. Budapest's broad avenues were littered with burned-out tanks and guns; buildings smoldered and burned, since firemen were unable to get through the improvised barricades or roadblocks of tanks; the pavements were littered with broken glass, debris, rubble, spent cartridges, and corpses. Nagy, who had been trying to strike a balance between the demands of the Russians and the Hungarian hard-liners on one side and those of the insurgents on the other, finally gave in to the insurgents and sacked most of the hard-liners from his cabinet, replacing them with more moderate figures.

On the morning of October 28 Nagy announced further concessions, to the consternation of Mikoyan, Suslov, and the Soviet ambassador to Hungary, Yuri Andropov: a general amnesty for all fighters, the disbanding of the AVH, and the replacement of the Soviet star on the national flag by the arms of Kossuth. Although the fighting still continued unabated, many other events were taking place: the United Nations was at last debating the Soviet Union's actions in Hungary; the Yugoslavs and the Poles expressed their support of Hungary; and all over Hungary student groups and insurgents were beginning to form what amounted to local democracy, and demanding Hungary's withdrawal from the Warsaw Pact and an immediate end to the Russian occupation.

It may be that Andropov, Mikoyan, and Suslov talked some sense into Khrushchev, and by the night of October 28 there were signs that the Russian troops were pulling out of Budapest, rather than being reinforced. In Moscow, the Central Committee was in a state of confusion, divided between those who wanted to support the new Hungarian government and those who wanted to use force against it. Khrushchev cannily observed,

"Politically, this is beneficial for us. The English and French are in a real mess in Egypt. We shouldn't get caught in the same company." All this of course was on the condition that Hungary remain within the Warsaw Pact, which was rapidly becoming a major—and critical—issue.[3] On October 29 it became clear that the insurgents had won, in part because of Moscow's hesitation to reinforce the Soviet army. Long lines of Soviet tanks left Budapest and moved south or west, leaving their dead behind in the streets; an eerie quiet reigned over the city; streets that had been named after Stalin and other communist figures were hastily renamed; and the files of the AVH were thrown open, the torture cells exposed, and the secret files and reports thrown out into the street in huge piles and set on fire.

Jovial Soviet diplomatic spokesmen all over the world reported that the crisis was over, that Hungary now had a government with which they could negotiate, that "the firemen had put the fire out," and that it was now time to rebuild on new foundations.

There is an old Russian saying that the bear is never more dangerous than when he looks friendly, and this would swiftly prove to be true.[4]

Erich Lessing/Magnum Photos

Brief victory! Freedom Fighters on a captured armored car chat with two young women. Note the Hungarian flag with the Soviet emblem cut out of the center, and the Kossuth coat of arms roughly painted on the side of the vehicle.

6.

"Long Live Free, Democratic, and Independent Hungary!"

During all this period the news from Hungary produced an unusual degree of ferment nearly everywhere, and nowhere more than among young people. At Magdalen College, Oxford, however, where I was then an undergraduate, there was less excitement than one might have expected. In the first place, there was a long-standing tradition of British phlegm—it was part of upper-class tradition not to appear to be too excited about anything. Then too, while eastern European university students have an ancient tradition of revolutionary ferment, both on the left and on the right, Oxford undergraduates do not as a rule see themselves as manning the barricades, tommy gun in hand, or seizing the BBC. Finally, the Suez crisis dominated conversation—it was the great divider between left and right, even dividing some conservatives from their own party. In addition, most of the undergraduates were in the reserves, and expected to be recalled to serve in Egypt if things there got sticky. Not a few of my friends were discreetly checking to make sure their uniforms and kit were in good order, and even I had performed a quick mental inventory just in case a telegram calling me back arrived from R.A.F. Uxbridge.

As the days went by, however, my mind turned more and more toward Hungary. Not that I suffered from any kind of nostalgia for a place where I had never been, and about which I had not until now experienced any curiosity; but there was no denying that if you wanted to be where the action was, Budapest was the place, not Uxbridge or even Oxford.

I always felt sorry that I had missed the Spanish Civil War—I was only three when it broke out—perhaps because it played such a large role in the lives of my father's friends, starting with Robert Capa, the war photographer. My father himself had gone to Spain—not to fight, but to get several Hungarian friends in the International Brigades out when things turned bad; and he knew a great many of the Hungarians who had fought there, including some who returned to Hungary, rather than to London, New York, or Los Angeles, and later became members of the communist party. Of course I had read Hemingway, Orwell, and all the usual literature on the war—at twenty-three, I was exactly the right age for that kind of adventure, and here it was, 1,000 miles to the east of Madrid, presenting itself again twenty years later. I was determined not to miss out on it.

Michael Maude, who had gone with me to East Germany, was in readiness for a call-up back to Wellington barracks and the Foot Guards and declined to go with me; and at first the only people who shared my enthusiasm for the idea of going to Budapest were two American friends: Jaquelin T. Robertson and Harry Joe Brown, Jr., known to everyone as Coco.

Jack Robertson was a Virginian, wealthy, talented (he would later become a distinguished architect), and good-looking. He was the son of the assistant secretary of state and "old China hand" Walter Robertson. A Yale man, Jack was at Magdalen as a Rhodes scholar; as an admirer of Generalissmo Chiang Kai-shek ("the Gitmo," as Jack always referred to him), General Douglas MacArthur, Henry Luce, and John Foster Dulles, he felt it to be his place to serve on the barricades against communism. Coco Brown, the son of a movie producer on the West Coast, was the best friend of my old schoolmate Warner LeRoy. Coco had arrived in Oxford from Stanford and New York City, and moved into Jack's rooms, where he slept on a sofa in Jack's sitting room. He was so naturally charming, sweet-tempered, good-humored, and well dressed that everybody, including the head porter and the dons, assumed he belonged there; but in fact he was a member of neither the university nor Magdalen College—he was simply a friend of Jack's, who had drifted across the Atlantic in his wake. Occasionally he wore a borrowed academic gown, just to get into the spirit of things. He seemed exactly the right person to go to Budapest, a figure straight out of the pages of *The Sun also Rises*. He seemed to think so too—anyway, where Jack went, Coco went too, in his easygoing way.

We talked about it in Bond's Room, off the stair to the JCS, where the college elite met for drinks, and over a sherry at eleven o'clock in the morning we decided it was our duty to take part. After a couple more of Bond's sherries, neither of them saw any problem in going to Budapest. Coco planned to pick up a professional sixteen-millimeter movie camera to shoot footage for a documentary film. I should bring my Leica. They had a few things to do in London first, then they would drive via Paris to Vienna and meet me there at the Sacher Hotel.

The rumor that I might be going to Budapest had spread by then, and I soon had three new people anxious to go, all Brits, none of them as well known to me as Jack and Coco. Russell Taylor was a large, quiet man, of a size and build likely to impress even the Russians; Christopher Lord was wiry, very handsome, and debonair; and Roger Cooper was tall and shrewd, with a wicked sense of humor—and, an important point, the owner of a car, a brown Volkswagen convertible, aged and rusting but, he assured us, absolutely up for the trip. We had a few drinks in front of the fireplace in Bond's Room, and as the decanter went around between us we rapidly formed a close friendship.

Taylor and Lord had done their national service in the army, I recall, Cooper may have done his in the Royal Navy. None of us doubted that our combined military, air force, and naval knowledge would come in handy to the Hungarians.

We agreed to meet in London, drive to Vienna, meet up with Jack and Coco there, then cross the border into Hungary. Since the newspapers were full of desperate appeals from the Hungarians for medical supplies, we decided that we should carry as much as we could; and my aunt Alexa, now a very wealthy and merry widow indeed, agreed to pay for the supplies, while our family doctor, Tibor Csató, a Hungarian himself, wrote the necessary prescriptions.

When these boxes arrived from Boots Chemists my father stared at them with undisguised pessimism and sighed. "My poor boy," he said, "better you bring them food." He was glum about the whole thing, partly because he suspected that Alexa, whom by now he disliked, had influenced me to go; and partly because he had lived through two Hungarian revolutions himself—the red one of Béla Kun and the white (counterrevolutionary) one of Admiral Horthy—and had not enjoyed either experience.

He quite liked my companions, though. As he watched us pack up the car he remarked that we looked a bit like Balkan bandits. He had a point, I thought, as I stood looking at them on the pavement. Christopher Lord wore a "British Warm," suede chukka boots, and a high astrakhan fur hat like a cossack's; Russell Taylor a heavy fur-lined overcoat rather like that of the bad guy who seduces Lara in *Doctor Zhivago;* and Roger Cooper a naval duffel coat and Russian fur hat. I wore my R.A.F. sheepskin-lined leather flying jacket, a beret, corduroy trousers, and waterproof leather boots—a look made popular by Monty in World War II. My father went back inside and came out with two bottles of his best cognac. "You will need this," he said, handing them to me together with a thick wad of fifty-dollar bills, and stood on the pavement puffing his pipe as we puttered off in the VW toward King's Road.

On my way home I had stopped not only at Alexa's house in Kensington Palace Gardens to say good-bye, but also at the Ritz Hotel bar on Piccadilly to meet Graham Greene, an old friend of my family's who was anxious to give me a few words of encouragement. Graham and my father had spent a long time together in Vienna, making *The Third Man,* and my father looked to Graham for advice about my education, with the result that Graham, who liked young people, eventually became something of a mentor to me. Unlike my father, Graham was enthusiastic about my going to Budapest. The one subject on which he agreed with Winston Churchill was that young men should experience and enjoy the sound of bullets (it hardly mattered whose) whistling past their ears. He had invited a chap from MI6—Graham had worked for "the firm" in West Africa during the war, and still maintained a shadowy connection with it—who wanted to meet me to join us, he said. Eventually, a tall, military-looking gentleman of a certain age in a well-cut suit and a Brigade of Guards tie turned up, and sat down without giving his name. Was I taking a camera to Budapest, he wanted to know? I said I was taking my Leica, yes—the influence of Bob Capa on my childhood was still strong. Good, the man from the firm said, just the ticket—I should photograph as many unit markings on Soviet vehicles as I could. This was the bread-and-butter of intelligence work—knowing which units were where, and what they were armed with.

My experience behind the Iron Curtain had already taught me that there was hardly anything the communists liked less than a camera in the hands of a Westerner; still, I saw no harm in agreeing. But how was I to get the film back, I asked?

The man from MI6 leaned close to me, and in a conspiratorial voice said, "French letters, old boy. Put two film cartridges in a french letter, tie up the end, then stick the whole thing up your bottom. Simple as pie." He sipped his martini for a moment. "Helps to lubricate it a bit first, you know. Vaseline, that kind of thing."

Graham nodded, either because he was already familiar with this piece of tradecraft, or because he had just memorized it to use in one of his novels.

The man from the firm congratulated me on the idea of taking in medical supplies—perfect cover! I pointed out that they weren't really cover—that was the purpose of our journey. He beamed. Better still! The best cover was always genuine. It occurred to me, belatedly, that he knew a surprising amount about our plans.

His expression turning more serious, the man from the firm wrote out a telephone number on a cocktail napkin and handed it to me. "Just in case you run into trouble, call this number," he said. I was about to put it in my pocket, but he shook his head. I should memorize it, he advised. I stared at it for a while—long enough to commit it to memory, I hoped—and then handed it back. He unfolded the cocktail napkin, took out a silver cigarette lighter, and set fire to it, causing a certain amount of consternation at the other tables and among the staff. He carefully crushed the ashes in the ashtray and dusted off his fingers. I must use it only in a *real* emergency, mind, he warned me. "Life or death, eh?"

I nodded, not at all sure now that I remembered the number at all, and wondering what kind of life-or-death situation would give me a chance to make a telephone call.

Graham and I followed him out onto Piccadilly, where he turned right and set off at a brisk pace, swinging his umbrella, a neat figure in a belted Burberry and a bowler hat. He was on his way back to the office, I guessed, the anonymous redbrick Victorian building on Shaftesbury Avenue that housed MI6 and was supposed to be top secret, except that London taxi drivers invariably pointed it out to tourists.

"You'll be all right, now" Graham said, with complete conviction. "I told your father—they always look after their own." The only other visit I paid was to say good-bye to Tracy Reed, Carol Reed's stepdaughter, in King's Road. I had strong feelings about her, which were not, I regret to say, fully reciprocated; but I discovered, as so many have before me, that every woman has a soft spot for an anticipatory hero on his way to war.

All that night, we drove hard, and in extreme discomfort, crossing to Belgium on the ferry and driving in shifts from there southwest. Between the bulk of Russell Taylor and the medical supplies, the little car was as crowded as a midget submarine; and like all Volkswagens of that era it was stiffly sprung and noisy. Thanks to the combined genius of Albert Speer and Dr. Porsche the gas stations on the autobahn were spaced at intervals determined by the range of a Volkswagen with a full tank—the car and the highways were, in fact, a single system, which is why early VWs had no gas gauge—and we stopped at each station, filled up, had a cup of coffee and a Steinhäger, and then drove on numbly through the rain.

In the early evening of October 29 we pulled up at the Sacher Hotel and stumbled into the lobby, cramped, stiff, and hungover. The concierge recognized me from the days when I had stayed here with my father when he was making *The Third Man,* over eight years ago; but then a concierge, I reminded myself, is paid—or tipped—precisely for his ability to remember faces. His expression was apologetic. It was a pleasure to welcome Herr Vincent Korda's son back to the Sacher, but he regretted that he had received a message from my friends, who were still in Paris and unable to join me. He read from the message pad: I should go on to Hungary without them.

I was not flabbergasted. Coco was mercurial. Once he reached Paris, the idea of going on to Budapest might easily seem to him less attractive—Paris has that effect on Americans. And I thought it very likely that the moment Jack's father heard about the idea, he had put his foot down. The idea of the son of the American assistant secretary of state being tried or jailed in Hungary was simply unthinkable—just the kind of embarrassment that President Eisenhower didn't want or need. I had been looking forward to a two-car convoy consisting of Roger's VW and Jack's more comfortable MG sedan, but we would just have to carry on, squeezed in like sardines.

The concierge expressed concern. Was Herr Korda truly planning to go to Budapest? I said I was, yes. Normally, of course, he said apologetically, he would call ahead and make a hotel reservation for us there—the Gellért Hotel was not too bad, though not like the old days, of course—but at present the telephone lines were unreliable. I briefly wondered which "old days" he was referring to, but I suppressed the thought. He had heard that some of the best hotels had been damaged in the fighting. If all else failed, we should try the old Astoria Hotel, which was now called the Red Star, and which used to be very comfortable. As for restaurants . . . He shrugged. Who could say? We must expect things to be very bad there, from the *point de vue gastronomique*. I explained that we were on a mission of mercy, rather than tourism or gastronomy. "*Ach so,*" he said briskly, snapping his fingers for the porters to take our luggage.

We ate a stupendous meal of soup, Wiener schnitzel, and the famous Sacher torte—after all, who knew when we would dine like this again?—drank several bottles of wine, and spent a quiet night. In the morning, after an excellent breakfast, we followed my father's advice and bought a couple of big hampers of delicatessen—salamis, smoked hams, sausages, cheeses, kirsch, and schnopps. It was my father's firm belief that in wars and revolutions you needed as much food as you could carry, both to eat and to give away in emergencies; and, as usual, he was right.

At the last moment, as we were trying to pack all this into the car, the concierge rushed out, cradling something in his arms. It turned out to be the large Union Jack that had been flying from above the hotel entrance, alongside the Stars and Stripes, the French flag, and the Austrian flag. With the help of the doorman, he tied it over the roof of the VW. It covered practically the whole car. "In case of strafing," he explained, pointing toward the cold, gray sky, as if Russian fighter planes might appear there at any moment; when I offered to tip him, he held up his hand with great dignity, perhaps the first time that the concierge of a major hotel has ever refused a tip. "Good luck," he said formally. "Bring it back when you return." He, the doorman, and the bellboy stood to attention and saluted as we left, puttering down the Ringstrasse in search of the highway to Budapest, feeling, at least in my case, every inch the hero.

Journey to a Revolution

The road between Vienna and Budapest is not one to inspire awe, especially in the early morning of a cold, wet, late-autumn day. At this point on the map, Hitler and Speer apparently lost interest in building autobahns, so the road was narrow, slippery, and lined with bare trees, and the countryside flat and shrouded in mist. The heater in the VW had given up the ghost—even when working, it was either hot as a furnace or freezing cold, nothing in between—so we kept ourselves warm with a couple of bottles of kirsch. There was not much in the way of traffic—in fact, for long periods of time ours was the only car on the road. Eventually, we arrived at the Austrian border post, where the guards stared at our passports incredulously. "You're going *into* Hungary?" the senior one asked.

We told him we were. Was that a problem? He rolled his

eyes. "Not for me," he said. "For you, maybe. Bad things are happening over there, you know."

"That's why we're going."

He shook his head. "A lot of people are trying to get out, you know. Not many people want go in." He stamped our passports neatly—the tradition of exactitude in the Austrian civil service never dies, as if Emperor Franz Josef's portrait were still hanging on the wall behind the desk—and handed them back to us. "In my opinion," he said gravely, "the world would be a better place if people stayed in their own country and minded their own business, *na*? *Gute Reise, meine Herren.*" He waved his hand, and a guard raised the red-and-white pole to let us through. They all stood and stared at us as we drove down the road, no doubt saying to each other, "All the English are crazy."

The border had been heavily fortified with tank traps and barbed wire on the Hungarian side—this, after all, was the Iron Curtain, right here, where it was not a metaphor but real—but it was now unguarded. No border guards or troops were in sight. We pulled into the guard post, the roof of which was festooned with Hungarian flags from which the Soviet star had been cut out with blunt scissors or a knife, leaving a ragged hole in the middle. Inside the guard post a man sat behind the desk, wearing a leather jacket, a green uniform cap from which the badge had been torn, and a three-day growth of beard. On the desk in front of him was a bottle with no label, the contents of which looked suspiciously like gasoline. On the other hand, it might not be gasoline—he looked drunk.

He stared out the window moodily at the Union Jack tied to the roof of the car. "More English journalists?" he asked. "We already have enough."

"We're English, yes," I said, holding out our passports, "but not journalists. We've come to deliver medical supplies to the hospital in Budapest."

He raised an eyebrow unsteadily. "No kidding? American cigarettes would be better." He flicked through our passports. "Korda? The movie producer is your dad?"

"My uncle."

"Son of a bitch! This calls for a drink." He uncorked the bottle, wiped the neck on his sleeve, and passed it to us. I took a swig. It tasted a bit like gasoline, but with a sweet, fruity aftertaste, and produced instant paralysis of the lips and unsteadiness of the legs.

"*Barack,*" he explained. "Homemade peach brandy." After each of us had taken a swallow he took the bottle back, took a heroic gulp from it, and put the cork back in. "It's the national drink," he explained. We expressed our admiration for it, so he passed the bottle around again. He had trouble with his passport stamp, partly because his hands were shaking, partly because somebody had cut the communist emblems out of it with a penknife, so it left a gluey mess of ink on our passports. "There's supposed to be a fee, but fuck it." He drew a thick line across the visa with a blunt pencil, and wrote "Gratis."

As we went back to the car, he called out, "English! Go careful, yes?"

"Careful of the Russians? Quite."

"Fuck the Russians—they're miles away. They're going home. Be careful of everybody. When you go through Győr be extra careful—there are some very bad guys in Győr."

He was wrong about the Russians. A couple of miles down the road we came around a bend and braked hard. Coming directly toward us, right down the middle of the road, was a long file of

Russian T-55 tanks. They did not slow down, so we pulled off the road and got out. As they started to pass us, thundering and belching smoke, I took out my Leica and pointed it at the closest tank—a perfect opportunity to get a photograph of unit flashes for MI6, I thought. Before I could take the picture, the gunner, leaning out of his hatch on top of the turret, slowly cocked his big 12.7-millimeter machine gun—even above the noise of the tanks I could hear the distinct *click-clack*—and swung the gun to aim directly at me, as the whole column ground to a halt. He had a leather helmet with earflaps, and the blackest, thickest eyebrows I had ever seen on a person. The eyes beneath them were flinty.

Very slowly I let the camera down on my chest and held my hands up. Right then and there I decided that MI6 was going to have to do without my photographs.

An officer appeared from a Russian jeep and strode over to us, his hand on his pistol holster. He wore a long greatcoat, down to the heels of his boots, a Sam Browne belt, and gold shoulder boards. His face was fixed in a scowl. He glanced at our Union Jack with distaste. "*Angliskiye*?" he asked.

We nodded and handed him our passports, which he thumbed through without taking off his gloves.

"*Vy kuda*?" ("Where are you going?") he asked.

"To Budapest," I explained.

He raised an eyebrow. "You speak Russian?"

I nodded.

"*Pochemu*?" ("Why?")

I did not think it was a good moment to bring up the subject of the Royal Air Force language course. "To read Pushkin in the original," I said.

He handed back the passports. "Fuck Pushkin." He looked into the car and saw the boxes. "What's in those?"

Medical supplies for the Budapest hospitals, we explained. We were on a humanitarian mission.

He raised an eyebrow. "You're bringing medical supplies to these sons of bitches? Fuck their mothers."

He pointed his finger at me. "You, the Pushkin lover. No more fucking photographs, you understand?" Then, in a more friendly tone: "Be careful going through Győr, English. There are very bad people there."

He waved us on, and went back to the jeep. The machine gunner grinned and gave us the finger as we drove past him on the verge of the road.

We briefly contemplated not going through Győr, but a look at the road map the concierge at Sacher's had given us showed that this was almost impossible. Besides, we needed gas, and of course the Hungarian road system, such as it was, had never been designed with civilian traffic in mind. Since there were hardly any private cars, there was no need for gas stations, and the few that existed were in the cities. Going through Győr seemed a better bet than trying to get around it on doubtful roads and running out of gas in the middle of nowhere.

Győr, when we got to it, was singularly drab and depressing, an eastern European industrial city, grim, gray, sooty, and apparently deserted. The broad, straight avenues* so typical of Hapsburg city planning were empty, except for a few burned-out trams and trucks; the shops were shuttered. Small bunches of flowers or a patch of white lime marked places where people

*Such avenues and boulevards had first been developed by Baron Haussmann for Napoléon III, during the rebuilding of Paris, because they offered a good field of fire for artillery in case of revolutions; other European monarchs quickly imitated them. The Russians took full advantage of this example of city planning in 1956.

had been shot, but the bodies had been carried away. It was lunchtime, and after a morning of drinking kirsch and *barack* we were hungry, so we pulled up at what seemed to be the only hotel, in front of which were parked a lot of dusty Russian-made cars, and went in. The desk was empty, but to the right of it was the entrance to a kind of *Bierstube,* part bar, part restaurant, in a kind of kitschy countrified German style—wood paneling, simple furniture, antlers, low ceilings. It was very crowded, and everybody was smoking Hungarian cigarettes, so there was a kind of low fog of vile-smelling smoke, coupled with the strong smells of damp wool, peach brandy, and paprika. The people in the room stopped what they were doing and stared at us, and it came to my attention that many of them were armed. Most of the men were swarthy, robust fellows, wearing shiny ankle-length black leather trench coats, with a submachine gun hanging from the back of their chairs, or a pistol on the table in front of them. The women were blond, heavily made up, wore black berets, and showed a lot of leg in high heels and stockings. They did not look like student revolutionaries, or exude a feeling of idealism.

"What do they look like to you?" Christopher asked me.

"Thugs?"

"Quite. I imagine they're AVH men and their girlfriends running for their lives. I suppose they're the bad guys everybody was warning us about."

I supposed so too, but there was no point in backing out, so we sat down at the only empty table and ordered beers. Russell asked to see a menu, but that request produced a round of laughter, once it had been translated. There was no choice of food. A stout waitress in peasant costume brought us each a bowl of goulash and a huge chunk of black bread from the kitchen. The goulash wasn't bad, either—hot, spicy, and filling.

One of the men in black leather got up, slung his tommy gun over his shoulder, and came creaking over to our table. "Americans?" he asked.

We explained that we were English. His expression registered disappointment. In those days people wanted dollars, not pounds.

"Journalists? We don't need no more journalists."

We explained about our mission of mercy.

He was incredulous. "You're going to give this stuff away? You could sell it for a lot of money in Budapest. Or here."

"We're not selling, we're buying. We need gas."

A crafty look came over his face. "You have dollars?"

I thought for a moment. It took only the smallest amount of survival instinct to tell me that revealing we had dollars to a roomful of fleeing AVH officers and their mistresses was like begging to be robbed at gunpoint and perhaps shot. I went out to the car, brought back my father's two bottles of brandy, and put them on the table. Our friend picked them up and looked at the labels. "French? Ten liters of gas."

"Twenty."

He shrugged. "OK." He went out and came back with two of the ubiquitous jerricans that the Germans had used for carrying fuel during the war. That was war—and revolution—for you. The bottles had probably cost ten pounds apiece at Justerini and Brooks, and we had just exchanged them for twenty liters of gas. When things get bad enough men will give a gold watch for a loaf of bread, and women their virtue. War and revolution teach you the relativity of values pretty quickly.

Christopher opened up the jerricans and sniffed at the contents. "They could be full of water," he whispered, revealing a hitherto unsuspected shrewdness that was to pay dividends in the days ahead.

We paid for our meal with Austrian schillings, and then I went in search of a toilet before we set out for Budapest. The hotel was empty—not a neutral emptiness so much as a threatening one, a feeling with which I was to become familiar in the next few days. The silence here was like the breath we all draw before something we know will be painful happens—the dentist's drill, the awful moment of truth, the act of unkindness that can't be put off a moment longer. The hotel was like that, its empty corridors more frightening than a roomful of armed men on the run. Where had the staff gone, I wondered. I opened a couple of doors, but they were closets—I should have asked what the Hungarian was for "men's room"—then saw one that had a crude painting of a prancing stallion on it, virtually a universal symbol from here to Las Vegas.

The door was stiff, but I pushed it hard. It was the men's toilet all right; I could see a sink and a urinal in the dim light from a partly opened window. I pushed harder, but the door was stuck. Then I saw why. My eyes quickly took in the soles of a pair of heavy, well-polished boots, the kind of thing tourists used to wear for hiking, a leather coat, and a gloved hand. There appeared to be blood on the tiled floor—not a lot, but enough. I was reluctant to step in it. I went to the next door down—with a painting of a horse with a long, flowing mane, and doubtless the Hungarian word for "fillies"—and went there instead. The toilet was blocked and there was filthy water on the floor, but at least there was no corpse. It occured to me then and since that I might have imagined the whole thing, but I was not tempted to go back and check.

I decided not to tell the others what I had seen—it could

only diminish their spirits—but although I had given up smoking a year or so ago, I took a cigarette from Christopher and lit up.

A few hundred yards down from the hotel was the government-owned gas station. A few men in black leather coats, like our friends from the pub, were filling jerricans from the one pump, some without taking the lit cigarette out of their mouth—with so many ways to die, ordinary risks don't seem to matter anymore. There was a body lying in the street, not far from the gas pump.

I took a long drag on my cigarette—I would smoke like a chimney for the rest of our time in Hungary—and thanked the stars that we had thought to pack the odd spaces in the car with cartons of Senior Service.

Suddenly we seemed to be a long way from home.

That was October 30. Though we didn't know it, events were moving rapidly even as we approached Budapest.[1] The day before, the Russians had agreed to withdraw their troops from Budapest on the condition that a government in which they had confidence was in power. No less a personage than Marshal Zhukov, the conqueror of Berlin, had announced in Moscow that the Soviet Union had confidence in Nagy's government, and throughout October 29 units of the Red Army began moving out of the capital and heading east, toward the Soviet Union, to demonstrate this. (Of course, we had seen Soviet tanks moving west, toward the border, but we were in no position to tell anybody as we drove toward Budapest.)

On October 30 Nagy declared the end of the one-party state, reestablished democratic government based on elections, and permitted noncommunist political parties to resume their existence. Even more dramatically, Nagy released Cardinal József Mindszenty from the prison where he had been held since 1948. The cardinal returned to his archdiocese in Buda amid scenes of enormous emotion, blessing the crowds from an armored car.

In the meantime, throughout October 30, widespread massacres of AVH officers and men were taking place—the photographer John Sadovy of *Life* captured one such incident outside the headquarters of the communist party of Budapest, which ended with dozens shot at point-blank range and the half-naked body of one of the officers strung up from a tree, doused with gasoline, and set on fire. Similar scenes of violence were taking place all over the city, as Hungarians unleashed their rage against the hated communist secret police. In the prime minister's office Nagy pleaded for an end to the violence, well aware of how it was infuriating Mikoyan, Suslov, and the Soviet generals, but there was no stopping it. Nearly a decade of political cruelty, repression, and murder was bearing fruit.

Meanwhile, in the Middle East, the Israelis had already attacked the Egyptians in the Sinai—the pretext for the coming Anglo-French assault—and in Cyprus the British and French paratroopers were already assembled, waiting for the order to go, while in eastern Hungary the first units of the Red Army were crossing the Hungarian frontier in huge numbers.

Unconscious of any of these events, we arrived in Budapest that afternoon, Roger at the wheel, Russell sitting beside him, and Christopher and I in the backseat, with a street map of the unfamiliar city unfolded on our laps. Our Union Jack produced sporadic applause from people on the street, who clearly supposed we were the forerunners of a vast Anglo-American force.

As we made our way through the city—which resembles a hillier Paris, divided by the Danube instead of the Seine, but with the same huge, ornate public buildings, Third Republic apartment buildings, and wide tree-lined boulevards—toward the center, the signs of fighting became more evident. On all sides there were blackened and burned-out tanks and artillery pieces. Many bodies still lay where they had fallen, sprinkled with lime if they were Russian, their heavy greatcoats spread out as if they were making snow angels, or covered with small bouquets of flowers if they were Hungarian. The corpses of women were particularly sad, skirts hiked higher than they would have been in life, slips and stockings showing, as if their privacy had somehow been invaded by death. Their handbags or string shopping bags often lay beside them—proof that the determination to stand in line for bread and groceries even when fighting was going on, so often praised by Western newsmen, was sometimes fatal. The litter of death and destruction was scattered all over the streets and pavements around them, much of it paper—the dead always appear to be carrying more paper than the living—identity cards, letters, family photographs, and shopping lists mixed with the burned and blackened bits and pieces from the bonfires of communist books, lithographs of Stalin and Rákosi, and communist party newspapers.

One had to drive a slalom course around improvised barricades of trams and buses shredded by artillery, interlaced with torn, twisted tram rails and ornamental fencing from the parks. The streets were littered with broken glass, debris, rubble, and spent cartridge cases—one proceeded slowly to avoid driving over a corpse, or bursting a tire on shards of armor from damaged tanks or unexploded artillery shells. Some of the tanks were still burning, filling the air with dense diesel fumes and the smell of charred human bodies. If one looked—though one tried hard not to look—the crisp, blackened bodies of tank crews and gunners hung out of open hatches, clawlike hands still reaching for safety, or perhaps mercy. The air was cold and almost unbreathable: a mixture of brick dust, cordite, and the stench of death that brought tears to one's eyes. Many people on the street held a handkerchief to their nose, and with good reason—we were soon coughing, hacking, and chain-smoking ourselves.

Here and there a tank had brewed, and the resulting explosion, as the ammunition stored in the turret went off, had blown the turret and the gun fifty yards or more away from the hull—something to give one pause, considering that a T-55 tank turret and its gun probably weighed ten or fifteen tons, and contained the tank commander, the gunner, and the loader. The turrets lay on the cobblestones like huge crushed beetles. At first glance all this seemed to indicate a splendid victory for the insurgents, but of course a certain number of the tanks and artillery pieces were no doubt Hungarian—losses had been high on both sides—and there was a singular absence of jubilation among the people on the street.

From time to time we stopped to ask directions for the Second Medical Clinic—Tibor Csató had instructed us to deliver the medical supplies to an old colleague, Professor Hajnal, director of the clinic. Almost everybody offered us a drink while explaining how to get there, and we felt obliged, out of politeness, to accept. We drank an awful lot of *barack,* and quite a few things that tasted even stranger than that. Most people asked if there were more of us on the way, a question we answered as evasively as we could. Our popularity was such that we felt vaguely guilty. We were handed flowers, bottles of liquor, souvenirs—

Jack Esten / Getty Images

Street scene, central Budapest.

Soviet cap badges and metal fragments from Stalin's statue—and pieces of paper with the names and addresses of relatives in Cleveland or Chicago. We passed back English cigarettes and salami and ham from our hamper. By the time we drew up under the porte cochere of the hospital, a vast grim pile in the most florid nineteenth-century Hapsburg municipal style, we had gathered a crowd of well-wishers, who explained our mission to the hospital staff. A stern but attractive young woman in a white coat splashed with blood agreed to take us to the director.

The elevators were not working, so, carrying our cartons, we followed her up the enormous ornamental marble stairway to the second floor, and through a pair of double doors into a long corridor lit only by daylight filtering through gray frosted windows. Here, there was plenty to attract the eye, until you learned not to look. The hospital was so full that many surgical cases had been left in the corridor, to be cared for by the patient nurses and the exhausted medical students. Although there was a shortage of anesthetics and painkillers, the wounded were mostly quiet. The wounded children were the hardest to look at, their big eyes following us as we trudged down the corridor past their beds.

Our guide knocked on a big door and showed us into the director's office. It was big—two walls were filled with medical books and one with framed diplomas. Professor Hajnal sat behind a massive desk, a small, intense man, with deep rings under his eyes, wearing a starched white coat and a stethoscope around his neck. He rose when we came in, and shook our hands with enormous formality and dignity after we had explained how we came to be here. "My old friend Tibor Csató," he said with a smile. "But this is very kind of you, to come all this way, at a time like this!" He looked at our cartons. "This is the first help we have had from the West, you know. Some of my young people thought there would be big Red Cross convoys from Vienna, but I was less confident." He shrugged. His English was heavily accented, but quite good. "Unfortunately, we are on the Danube, not the Suez Canal, eh?"

We had brought along a typed inventory of what was in the cartons, prepared by Boots, and Professor Hajnal looked it over quickly. "Good," he said, "very good. We can use all this, thank you. We are short of everything." He went to the window and looked out, silent for a moment. It was growing dark already, and seemed all the darker for the absence of streetlamps. Electricity was a problem. In the distance there were occasional shots. He sighed, then turned around briskly to face us. "It's a nasty business out there, eh? And I think it's not over yet, whatever people say." He sat down at his desk and looked at us sternly for a moment. "Now you have delivered the medical supplies, what do you propose to do?"

Christopher supposed we would stay on and make ourselves useful.

Hajnal shook his head wearily. "Useful? Doing what, please?"

We had a car, Roger pointed out. We could help bring the wounded to the hospital. At the moment, we had a British flag on it, but perhaps we could change it for a Red Cross flag instead?

"No," the professor said. "You are much better off as you are. The Russians often shoot at ambulances, you know. They might not shoot at the British flag, who knows? Well, you're young, you'll find something to do. But let me give you this."

He put a sheet of paper on his blotter, unscrewed his fountain pen, and in a neat, clear hand that exactly resembled my father's (it must have been what was taught in Hungarian schools) wrote, on one side in Hungarian, on the other in English:

> To whom it may concern
>
> I hereby testify that Mr. R. Cooper, Mr. M. Korda, Mr. C. Lord and Mr. R. Taylor have today delivered to the IInd Medical Clinic (Budapest) 1000 million units of penicillin. We are very grateful to them for their generous present and for their courage in bringing it in these troubled times to Budapest, which help our patients and wounded.

He read it over, stamped it with an ink stamp and signed it on both sides, then folded it in four carefully, slipped it into an envelope, and gave it to me. "I hope it is useful," he said. "Unfortunately, I don't know Russian, but usually they have a great respect for official pieces of paper even if they can't read them. They love ink stamps, you know. Russia doesn't run on oil or coal—it runs on ink. Good luck to you."

The stern young woman led us back to our car. We asked for directions to the Red Star Hotel, on Kossuth Lajos Utca, and she did a little sketch for us in her notebook, tore the page out, and gave it to Christopher. "Be careful," she said.

"I thought the Russians had gone?" Christopher said. He had the kind of good looks that made all the young women we met talk to him as if he were our leader or spokesman. It was an interesting thing to watch, though a little depressing.

"The Russians, maybe, but there are barricades, patrols. People shoot first, ask questions later, yes?"

"Nice girl," Russell said as we bundled ourselves back up in the car. "Shame we didn't ask her to join us for dinner."

To whom it may concern

I hereby certify that Mr. R. Cooper, Mr. M. Korda, Mr. C. Lord and Mr. R. Taylor have toddy delivered to the IInd Medical Clinic (Budapest) 1000 million units of penicillin. We a very grateful to them for their generous present and for their courage in bringing it in there troubled times to Budapest which help our patients and wounded.

Budapest, November 3rd 1956.

(Professor) I. Haynal
Director of the IInd
Medical Clinic

The trip to the hotel in the dark was an alarming experience. There were indeed plenty of barricades, and also plenty of armed patrols manning them, and at every one we were flagged down and stopped. Civilians stuck their heads, and very often their weapons, into the car and demanded to see our papers, punctuating their requests by clicking the safety catch of their weapon off and on repeatedly. "Bloody poor fire discipline," Christopher remarked, with sound professional judgment—fire discipline was not the rebels' strong point, and from all over the city you could hear the occasional shot, or a short burst of submachine-gun fire, as they emphasized their demands for identification. It was not a lot of shooting, as these things go, but it was enough to keep your foot close to the brake pedal and your British passport in your hand. Our flag brought us a lot of offers to share a drink, which we felt obliged to accept, but it also brought a certain amount of hostility. Where were the British and the Americans? Why had they not supported the revolution? What the fuck were we doing in Egypt? We took to saying, "Plenty more of us on the way," in a cheerful tone of voice. "Keeps their morale up," Roger remarked.

Russell sat in the back reading the Baedecker's guide to Hungary I had taken from my father's bookshelves (and which is still in my possession) with the help of a flashlight. "The Astoria Hotel gets a good rating," he said. "The restaurant rates a star too. *Eszterházy rostelyos* is rather a specialty, it says here. Beef roasted with onions, tomatoes, and paprika—doesn't sound bad, does it? They do amazing things with lake fish too, apparently."

"What year is that guidebook?" Christopher asked.

Russell looked at the copyright page. "Nineteen twelve."

Eventually we found Kossuth Lajos Utca, and the hotel. The red stars had been taken down, but there had been no time to make new signs. It was large, in the European tradition of *hôtels de luxe,* but the grandeur was faded now, and the white-and-gold lobby, with its huge chandeliers, looked uncared for. The concierge took one look at us and shook his head. "We are full up with British journalists, *meine Herren*. I haven't a room to spare."

This was my specialty. I put one of my father's fifty-dollar bills inside my passport and handed it to him. He pocketed the bill and examined the passport, and his expression changed instantly. "Korda!" he said, giving the name its full Hungarian pronunciation. "Not the son of Sir Alexander?"

"His nephew. My father is Korda Vince."

"Ah, the *régisseur*. I admired so much his sets for *The Private Life of Henry VIII*. I had the honor of serving your uncle when he stayed in the hotel in 1918, with his beautiful wife Maria Corda. . . . They had the best suite in the hotel." He sighed, looking around the lobby. "Things aren't the same these days, of course. But I can give you gentlemen a comfortable room. For how long will you be staying?"

"We don't know yet."

"For some time, I think. I hear the border is closed."

"Is the restaurant open?" Russell asked.

"Alas, at present, no, but we can serve you sandwiches in the bar."

Under the communists, the Red Star Hotel had been the place where Western businessmen and journalists stayed, so the rooms were comfortable and clean, and the bathroom was impressive by any standards. Of course it had suited the government to put foreigners here, since, as we would be shown tomorrow, every room was wired for sound, and there was a big room in the basement, equipped with tape recorders, where the AVH could record every word that was spoken. An AVH guard at the door had discouraged Hungarians from entering and meeting foreigners.

We put our luggage in the suite and went downstairs to the bar, which was full and humming. A gypsy orchestra played. Once again, as in Győr, the room fell silent as we entered, but this time the people staring at us were unarmed and spoke English—they were all British journalists. We sat down and ordered drinks and sandwiches, and for a moment were treated like celebrities. We were, after all, news, not big news, but still worth a story. "Oxford Students Arrive in Budapest," "Film Producer's Nephew and Pals Join Budapest Rebels," "British Students in Race against Time to Deliver Medical Supplies"—we would get a brief and largely erroneous story in most of the dailies tomorrow morning, and then be eclipsed by larger events.

We were joined by an attractive woman and her husband. Martha Dalrymple* was a foreign correspondent for the Manchester *Guardian,* her husband Ian a reporter for the more sensational *Daily Express*[2]. Martha was small, shrewd, well informed, and kind; she wrote long, thoughtful stories full of facts and history. Ian was big, bluff, florid, and hearty, a Fleet Street man to the bone, interested only in the kind of scoops the *Express* and its competitors thrived on: "Brighton Bathing Beauty's Pauper Funeral" or "West End Sex Club Scandal—Peer Denies All." How they managed to get along together was hard to understand, but somehow they did; he would go on to a distinguished career and a knighthood, while she became a hugely successful writer of internationally best-selling murder mysteries, one of

* I have changed the names of journalists, diplomats, and Freedom Fighters to protect their privacy.

which would be set in the Hungarian Revolution, with characters very much like ourselves in it. No real writer ever wastes anything, as Graham Greene liked to say.

Once Ian had pigeonholed us as a small story, he could be friendly enough, though he was like all journalists: anything you unguardedly said in his hearing could—and would—be used against you. "Not bad sandwiches," he said. "Let's have the chap bring us another big plate of them, and a bottle of wine, shall we? Or rather, two bottles."

This seemed like a jolly good idea, and we sat back and chatted about events. Ian, naturally, knew the sensational side of things, or at any rate the things which might seem sensational to a reader of the *Express*—our brief fame, he told us, was about to be eclipsed by the arrival of Judy Cripps, the daughter of former labour chancellor of the Exchequer Sir Stafford Cripps (Churchill's bugbear, whom he usually called "Sir Stifford Crapps"). "Ex–Cabinet Minister's Beautiful Daughter Joins Freedom Fighters" would certainly trump our story, Ian thought.

"I say," Christopher protested, "I know Judy. She's not what I'd call a beauty."

Ian dismissed this with a wave of the hand. "She will be for the readers of the *Daily Express,*" he said. "We can't call her 'plain,' can we? Anyway, I hear that at Oxford she's described as a raven-haired beauty who gives wild parties."

"That's Sarah Rothschild."

"I bow to your superior knowledge."

Ian went off to the loo, stopping to say hello to fellow Fleet Streeters on his way back—Harry of the *Times,* Jack of the *Telegraph,* what's-his-name of the *Mail,* chap from the *Mirror,* old Bodkin from the *Sketch*—but when he sat down his rosy face was grave. I wondered if there had been a dead body in the gents', but apparently not. He lit a cigarette and took a deep puff. "I say, bad news on the wireless, chums. We and the Frogs have just issued an ultimatum to the Gyppos and the Jews. Stop fighting, or we come in to protect the canal. We're all at the wrong party. It looks very much as if the real show is going to be at Port Said, not here. We're going to be 'below the fold,' as the Yanks say." He gave us an appraising stare. "I shouldn't wonder if you chaps were court-martialed as deserters when you get home."

The same thing had occurred to us.

"I don't think we're at the wrong party at all," Martha said, but I could see that as a professional, she knew her husband was right.

The waiter arrived with the sandwiches and the wine, which wasn't at all bad. I talked to Martha Dalrymple, who seemed to know everything and everyone in Budapest. She thought that Nagy—whom of course she had interviewed—was drifting at the mercy of events, and would probably be replaced by Kádár, who had once been his nemesis. In her opinion, the Russians would make their move soon, but not before Mikoyan and Suslov had gone home to Moscow—that was the event to watch out for. She had talked to Colonel Pal Maleter, who told her that there were now more Russian troops and tanks in the country than before, and of higher quality, and that nobody in the government wanted to believe it. We should be very careful, she warned—things could get very rough. The news from London wouldn't help matters—the world's attention would be focused on Egypt, not on Hungary.

Rougher than before, I asked? Oh, yes, she said, much, much rougher. And then there was Suez. . . . She sighed. We didn't know what was going to happen here yet, but it was already yesterday's news in London.

We finished the wine, ordered another couple of bottles, then went upstairs to bed, and slept like so many logs.

In the morning, we met a "delegation" of students, bringing us the greetings of the Budapest student associations. They seemed to imagine that we in some way represented the British student movement, and that we were here as "observers" on behalf of hundreds of thousands of aroused British students.

We rather bashfully accepted their greetings—it would have been rude to tell them that we represented nobody but ourselves.

Camera Press/Retna

Besides, the "continental" idea of a student is very different from ours. The Hungarian students were organized into communist groups and cells, and when they turned against the regime they naturally retained their organizations—they simply reversed their ideology, as it were. They had been among the first to demand changes from the government, and also among the first to take up arms against it. Students had led the Hungarian rising of 1848 too—the militancy of student groups was a fact of life in European politics, as the French government would rediscover to its cost in 1968. Students were politically informed and active; they took themselves seriously and expected to be taken seriously—they had, in fact, just contributed heavily to bringing down a Stalinist government in Hungary and to defeating the Russian army in the streets of Budapest. Explaining to them that undergraduates at Oxford had no political influence at all and were taken seriously by nobody was difficult, and seemed, in the circumstances, a tactless exercise.

Nor, of course, could any of the four of us have spoken for students at "redbrick" universities. All we knew about was Oxford, and only the part of Oxford that consisted of upper-middle-income (and upper-middle-class) public school boys. It is possible that there were undergraduates at Oxford meeting to protest the Soviet attack on the students and workers of Budapest, or the Anglo-French threat to attack to Egypt, but nobody we knew was likely to be doing so; nor would Anthony Eden be sitting at his desk at 10 Downing Street worrying about what Oxford and Cambridge undergraduates would do if he bombed Egypt.

October 31 was a day of surprise announcements, at least three of them fateful. The first and most fateful one was that of Imre Nagy, who told a cheering audience that Hungary had demanded its release from the Warsaw Pact. Hardly anything that Nagy could have done was more certain to make the Russians decide on full-scale repression of the revolution. This we heard about in the streets, as our student friends took us for a tour of the city, and they were wildly enthusiastic about it. It was, in their view, the first real step toward a free and independent Hungary, perhaps as a neutral country, like Switzerland, with a role to play as a bridge between East and West.

This was, of course, to deny the reality of Hungary's geography. If the Poles had tried to leave the Warsaw Pact, there were half a million Russian soldiers in East Germany to keep them separated from the West—Poland could be attacked from East and West simultaneously, squeezed like the filling in a sandwich. But an independent Hungary would roll the Iron Curtain all the way back to the border of the Ukraine, and for the first time directly expose Soviet soil to the possibility of an attack by the West. Seen from the point of view of anybody standing in front of a map in the Kremlin, this must have seemed like a direct threat.

We did not hear until later that day the next piece of fateful news—that the British and the French had bombed Egyptian airfields and destroyed most of the Egyptian air force on the ground. In the Middle East, the other shoe had dropped at last, with a bang.

Only toward the end of the day did we learn the most sinister news of all—that Mikoyan and Suslov had left Budapest and flown back to Moscow for "consultations" with the Politburo.

Martha—who clearly knew her stuff—had told me the night before that this was "the event to watch out for," and here it was, arriving on schedule, or even a little before. We did not, of course, know that on November 2, in Moscow, the Central Committee would make the final decision to replace Nagy with Kádár, and to suppress the Hungarian Revolution by force.[3]

Erich Lessing/Magnum Photos

. . .

None of these portents concerned our new student friends—or perhaps the students merely wished to appear upbeat and confident to the representatives of British studenthood in their midst. They mostly spoke decent English, while we knew no Hungarian at all, so we knew only what they told us or translated for us. We were taken on foot to see where the most significant battles had taken place; we went to look at what remained of the statue of Stalin—two giant bronze boots standing on top of a scarred, defaced pink marble base. We went to Parliament and trooped through the building, which had the improvised, revolutionary atmosphere of the Smolny Institute in Petrograd in 1917, and briefly saw from a distance Imre Nagy, who was returning from making his fateful speech, his face aglow with the applause that had greeted it—a short, plump man, with the graceful walk that so many fat men have. He was wearing a neat bourgeois suit and highly polished shoes, and his eyes bulged slightly behind his pince-nez.* We went to the university, to the statues of Petőfi and Bem, and were shown the radio station and the former headquarters of the AVH, both of them heavily scarred by shells and bullets and surrounded by many bouquets of flowers on the pavement where insurgents had fallen. We were introduced to armed student groups at various barricades around town; and we came to realize that there was not just one armed insurgency of Freedom Fighters, but three distinctly different ones—the students and "intellectuals," the much more heavily armed factory workers, and a large part of the Hungarian army itself.

*I later learned that my father had called his friend Zoltán Kodály, the composer, and asked him to reach Nagy and have him look after us. I am still sorry that we were added to the poor man's burdens.

Mirrorpix

The boots of Stalin's statue.

Feelings between the three groups were not always comradely, and the factory workers were distinctly more rough-and-ready and threatening, often led by very exotic figures indeed, and more motivated than the army or the students by a brutal spirit of revenge.

Their aims were not exactly the same, either. The students were idealistic—they wanted Hungary to achieve socialism "with a human face," a kind of Sweden-on-the-Danube. The factory workers were practical—they wanted decent working conditions and pay, and an end to unrealistic Stakhanovite "norms" and to the party's control over daily life. The army above all wanted to get rid of the Russians once and for all—one of the things that pleased the army most was Nagy's decision (never put into effect) to change its Soviet-style uniforms for the "traditional" Hungarian ones.

This is not to say that many of these aims did not overlap, of course. Still, hardly anybody, least of all Imre Nagy, envisaged a full-scale return to capitalism and private ownership of land. Ten years of communism had changed people's mind-set. Some wanted Marxism without the errors of Leninism and Stalinism; some wanted a more humane form of communism without Russian domination and police terror; hardly anybody was as yet aware that the whole system was a sham and a failure, and would lead nowhere. (The notion that Hungary would eventually emerge, thirty years later, with a thriving stock exchange, the privatization of nationalized industries, a booming market in luxury co-op apartments overlooking the Danube in Budapest, and streets full of expensive shops and restaurants was not one that would have occurred to Nagy or the insurgents, or necessarily have appealed to them.)

In later years, and even at the time, much was made of the

Freedom Fighters' anti-Semitism, but it has to be said that we saw no sign of it. Admittedly, we did not know Hungarian, but nobody even alluded to the fact that many of the most hated of the Hungarian Stalinists were Jewish, and we neither saw nor heard a single anti-Semitic slogan or remark. This is not to deny that if you had gathered a few dozen Hungarians to do battle against the Red Army in 1956 some of them might be anti-Semites—anti-Semitism has deep roots everywhere in eastern Europe. But anti-Semitism was not what motivated those who fought the Russians and resisted the Hungarian communists. They were not "counterrevolutionaries" or "White terrorists" or "Horthy supporters" or "Jew-haters," as the Russians and Kádár's government later accused them of being. They had simply had enough of a government that depended on lies, police terror, and Russian support, and that imposed on its people a system of rigid ideology, grueling deprivation, and Orwellian denial of common sense and historical reality. The awfulness of the Hungarian communist government managed to bridge the differences between workers, students and intellectuals, and the army—no mean achievement!

That night we dined again with Martha and Ian, who were weary and depressed. They had met briefly with Cardinal Mindszenty, who was demanding full restitution of the church's property and the resumption of its "traditional" role in Hungarian life, demands which Nagy, a lifelong communist, could not have met even had he wanted to. From all over the country, reports were pouring in of Soviet troop and tank concentrations heading toward Budapest from all directions, and of sporadic and hopeless fighting in many cities.

The next day, November 1, was quiet in Budapest, but there was an edge of fear that was hard to ignore. Trams and buses were running again on some streets; bakeries were open for business, and long lines formed outside them; even the flower sellers were back in business again. The general strike was over—not officially; it simply petered out—and people who could get back to work did. Occasionally, Soviet airplanes flew low over the city, and almost everybody cheered, thinking that they were American or British, and that they signaled the arrival of paratroopers, or at least aid. Having been in the R.A.F. and learned aircraft recognition, I knew better, but I did not think it was necessary to tell anyone.

Had we but known, Nagy was busy protesting to the Soviet ambassador about these same flights, and about the fact that the major airports were in the hands of the Russians. Ambassador Andropov (confirming the old definition of an ambassador as "an honest man sent to lie abroad" for his country) explained that the airports were being protected by Soviet armor in order to secure the speedy withdrawal of Soviet citizens from Hungary, and that the aircraft overhead were transports carrying Soviet citizens home.

We attended the daily press briefing at the British legation, and were impressed by the number of journalists, and by the air of elegance of the legation itself. There was not much in the way of news, although those who had attended the rival briefing at the United States legation were able to inform us that the United States Treasury had started to sell sterling on a huge scale, with the result that the value of the pound had dropped precipitously, and that Britain's gold and dollar reserves were fast drying up. This was not big news to the Hungarians, but it revealed to those of us who knew anything about Britain's finances that President Eisenhower, only a few days away from the election, had decided to drop the guillotine blade on Eden's neck.

The legation staff did not look on us kindly. A senior official took us to one side and angrily asked us what the bloody hell we were doing here, and had we any idea how much trouble we were making? Christopher replied that we had as much right to be here as anybody else, and that it was his impression that Britain's diplomatic service existed to help British subjects abroad, not to rant at them.

That produced a very cold look. The best thing we could do for everybody's sake would be to get the hell out of the country, the official told us stiffly, except that it was too bloody late—the Russians had closed the borders as tight as a drum.

"I bet you didn't tell Judy Cripps that," Christopher said.

Slightly deflated, the official sighed, and said, "Well, I did, actually, but she wouldn't listen either. The apple doesn't fall far from the tree. . . . Listen, if worse comes to worst—and it bloody will—you chaps come here, on the double. You've no business being here, but you are British subjects."

And it was true, after all. Everywhere we went in Budapest, we were greeted as if we were the personal emissaries of Queen Elizabeth II. Nobody felt that way about the journalists—it was their job to be here—but we had come here voluntarily and at our own expense, and people were to a surprising degree impressed and comforted by that.

That evening, as we sat drinking, we heard another piece of bad news, though it meant little to us at the time. János Kádár, the tough, ambitious hard-liner who had been arrested and savagely tortured during Rajk's trial and had then joined the government of Imre Nagy, was rumored to have fled Budapest in a Soviet armored car and to have taken refuge in the Soviet Union. This news did not set off alarm bells in our heads, but it should have. Kádár was a tough guy, an unsentimental realist, a survivor of torture. If he had decided that the revolution had gone too far—or that the Russians were going to crush it—then real war was very likely in the immediate future.

We did not know that Kádár, before fleeing to the Russians, had promised to go out into the streets and fight Russian tanks with his bare hands if they attacked; we did not know that Nagy had told the Soviet ambassador that Hungary was leaving the Warsaw Pact and declaring its neutrality and had informed the United Nations of that fact by cable; and least of all could we have guessed that Nagy's cable to the UN would be ignored at UN headquarters in New York, where attention was concentrated on the fact that the Israelis had by now seized the Gaza strip and the Sinai.

November 2 was another day of nervous calm, as the city "normalized" itself, while the Nagy government desperately tried to send out messages of distress. This became increasingly difficult when the international telephone lines were cut, "accidentally" as some in the government claimed (or hoped), but the British journalists rightly assumed the worst—the Russians had cut the lines again.

November 3 was a day of amazing beauty—a clear cold sky that made one realize, perhaps for the first time, what a beautiful city Budapest is. The flower sellers in the streets were joined by vendors of roasted chestnuts, with the smell that marks every autumn and winter day in central Europe. Bars and espresso coffee shops were open; the pastry shops were open for the first time; the streets were full of strolling people, very few of whom were armed. In Moscow, Nikita Khrushchev remarked to the Yugoslav ambassador that "Stalin had cooked up a porridge in Hungary that the present Soviet leaders now had to eat."[4] The ambassador rightly concluded that a massive Soviet attack was about to take place.

I went to the Gellért Hotel, across the Danube, where my uncle Alex had lived for a time during World War I, and where there was a world-famous swimming pool that produced artificial waves. At the Gellért, the concierge, once he was informed of my identity, left his desk to shake hands with me. He too had had the honor of serving my dear uncle. Was I enjoying Budapest? I would see—things were getting back to normal now. It would be a pleasure for him to arrange a few sightseeing trips for me and my friends—he could arrange for a car and an English-speaking driver, of course; I should not hesitate. There was very much to see. He hoped my dear father was well?

I thanked him, secure in the knowledge that Alex had been such a world-class tipper that no concierge anywhere in the world was likely to forget him, and that my father's way with hotel personnel was also one of legendary generosity. My father had stayed in the Gellért in the 1930s, when his friend Nicholas Horthy, the son of the regent, had invited him back to Budapest to seek his advice about rebuilding the film studio. My father had not enjoyed Hungary under fascism, much as he liked Nicholas Horthy and the Gellért. My uncle Alex had been briefly imprisoned in the cellars of the Gellért during the White counterrevolution, and had been saved from torture and execution by the prompt intervention of his wife Maria Corda—then a famous movie star, and always a woman of great temperament with a taste for operatic scenes—and my uncle Zoltán Korda (known in the family as Zoli), who learned where Alex was in the elevator at the Gellért, when they heard one White officer say to another, "We've got that Jew bastard Korda in the cellar, and tonight we're going to torture and shoot him after dinner."

Thus the Gellért had played a significant role in the history of the Korda family, and it was with great curiosity that I visited it for the first time.

That night the streets were quiet—no traffic, few people, no shots at all: peace, it seemed, at last. We could not have known that Pal Maleter—just promoted to general and vice-secretary of defense—had gone to meet with the Soviet generals at their headquarters to arrange the final details of their withdrawal, and there he had been arrested at gunpoint by none other than Ivan Serov, who was the head of the KGB and a close friend of Khrushchev's. (Maleter was shot, though his "execution" would not be announced until 1958.)

Still less could we have known that during the night the Egyptians would sink ships at both ends of the Suez Canal, blocking it tightly to all commerce, and ensuring that the French and the British, despite their cold feet—which were getting rapidly colder as the franc and the pound fell—would feel obliged to go in and take control of the canal.

7.

Götterdämmerung on the Danube

I woke in the early hours before dawn to the noise of thunder. The electricity at the Astoria Hotel was spotty, and in any case I didn't want to wake my companions, so I used my flashlight to make my way to the bathroom, which was like a minefield, strewn with Russell's apparently bottomless toilet kit. Like Michael Maude's grandfather, he traveled with a vast array of monogrammed objects—ivory-backed hairbrushes, manicure set, sponge bag, razor and razor strop—an amazing amount for a man who always looked slightly disheveled. On my way back to bed, I pulled the thick curtain back to look at the storm. To my surprise it wasn't raining. I could see what looked like lightning flickering in the distance beyond the hills of Buda, except that it was reddish rather than white. The low rumble continued steadily, like a basso profundo drumbeat; then, all of a sudden, it was followed by a vastly more unpleasant and sharper sound, that of explosions, none of them very close for the moment, but instantly recognizable to anybody who had lived in London during the blitz. I had been sleepy, and expecting to get back to bed, but all of a sudden I was fully awake, and ice-cold, even in my dressing gown.

I looked at my watch. It was about five in the morning—just the right time for an artillery barrage to "soften things up" before an attack. That was the one thing you could rely on when it came to armies—they always stuck to the same routine. This meant that at first light we could expect troops and tanks in the streets outside, and not for show.

I went over and woke up Christopher, who was a light sleeper. Waking up Russell was like trying to wake up a hibernating bear. Christopher opened an eye. "What the hell bloody time is it?" he asked.

"Fivish."

"Bloody hell. What's up?"

"Artillery. They're shelling the city."

"I thought they were going home?"

"Yes, well, apparently they're not. Best get up and dressed, don't you think?"

"Bugger! It's bloody cold."

It was the morning of November 4. If we had been listening to the radio—and if we had understood Hungarian—we could have heard the premier, Imre Nagy, announcing to the Hungarian people that Soviet forces had begun an attack on Budapest "to overthrow . . . the legal government." Nagy added that "our troops were fighting"; but if that was the case, there was as yet no evidence of it, or perhaps the fighting was taking place across the river or in the suburbs. We dressed hurriedly and went downstairs to see what was happening. Strangely, the hotel was silent—either the British journalists were heavy sleepers, or like myself they had heard what they thought was distant thunder and then rolled over in bed and went back to sleep. Most of them had left their shoes outside their door—the Astoria was still operating normally, and the valet picked your shoes up for polishing every night. I noticed Martha Dalrymple's neat little pumps next to her husband's big brogans. Many people had left the "Do Not Disturb" sign (in Hungarian, German, English, and Russian) hanging from the doorknob, but the Russian gunners had felt no compunction about disturbing people.

Outside in the streets, people were beginning to appear in small groups, heavily armed and grim-faced. Whatever illusions had sustained the Hungarians until now—that the Americans or the United Nations would come to their rescue; that Nagy could make a deal with the Russians for independence or even neutrality for Hungary; that the Russians, having been beaten once in the streets of Budapest, would not come back for another try—they now faced reality. The United States was within a few hours of a presidential election, and President Eisenhower was deeply concerned that the Soviet Union might overreact to events in the Middle East and Hungary and start a preemptive nuclear war; the United Nations, unwilling to offend the Soviet Union, was more concerned with two white, colonialist powers (and Israel) ganging up on an Arab country. Nagy had either overplayed his hand or lost his nerve, or both; and the Russians, smarting from their defeat, were coming back to take their vengeance for it.

Kossuth Lajos Utca was exactly the kind of wide boulevard that Napoléon III had in mind, when he redesigned Paris, for the use of artillery to cut down the mob. Since it led to the Parliament building, like almost all of Budapest's *grands boulevards*, there was little doubt that the Russians would be coming down it fairly soon. Several groups of insurgents were working on this assumption as we came out of the hotel, building improvised barricades. We talked with a group of students and intellectuals—the armed factory workers were less friendly toward foreigners, and seldom included anyone who spoke English. This group contained a pretty blond girl with a fur hat and a Russian submachine gun, and was led, of all things, by a university professor named Attila who taught English literature. The blond young woman appeared to be his girlfriend.

"Tear up the cobblestones," Attila advised us.

"Tear up the cobblestones?" I looked down at them. They were not small ones, like the ones you see in France and Belgium, but

Getty Images

big hexagonal blocks of stone, neatly fitted together and apparently cemented into place. "What difference would that make to a tank?"

Attila sighed, and explained. The Russian tanks had metal treads, which were ideal for fighting in the mud or traveling on dirt roads, but which tended to slip and slide on paved streets. Cobblestones were wet and greasy to begin with, and if you piled up enough of them, they were difficult for a tank to climb over—it was likely to lose traction. At worst, pulling up the cobblestones would at least slow a tank down, so somebody might be able to run up behind it and throw a Molotov cocktail onto the engine vents, or get a shot at the side of the tank, where the armor was thinner, with an antitank gun.

I was impressed. The general view in the Royal Air Force was that tanks were somebody else's business, not ours, but Attila had received his military training in a country which had been fought over by tanks in 1944 and 1945, and in which the notion of young people fighting as militia was not as far-fetched as it might seem to us. For all that he looked like the scholarly type, Attila, like most Hungarians, had received years of sound, practical military training, from people who took these things seriously.

All the same, as I scuffed at the cobblestones with my shoes, I was doubtful. "How on earth will you pick them up?" I asked.

Attila looked impatient. "With our fingernails," he said testily. "Once you get the first one loose, the rest are easy."

But before he could put this to the test, a truck pulled up, piled high with spades, picks, and crowbars, on top of which sat half a dozen rough-looking factory workers, armed to the teeth. "Take as much as you need," said the man sitting beside the driver, apparently the group's leader. "We liberated the whole lot from the road workers' stores." He swung out of the cab to unlatch the tailgate. He wore a dashing gray felt hat, of the Borsalino type, and a *Feldgrau* uniform very much like that of a German NCO, with the badges removed—probably the uniform of the old Hungarian army, under Horthy. On one leg he wore a high black boot. From the knee down, the other was a roughly carved wooden leg, which came to an end with a thick pad, like a piece of furniture. Crisscrossed across his chest were bandoliers of ammunition, like those of a Mexican bandit of the Pancho Villa era; he carried a military rifle, and strapped to his leather belt were a holstered pistol and four German "potato masher" stick grenades. He spoke some German—*Miklósdeutsch*.

"Where are the Russians now?" Attila asked him.

"Every-fucking-where."

The rumble of artillery continued, as background noise. From time to time, a shell landed with a deafening crack within a few blocks of where we stood, sending up a geyser of smoke, dust, and debris. We made it our business not to duck—that was clearly the order of the day, here in the streets. Anyway, if the girl with the tommy gun wasn't going to duck, neither were we.

"Have they reached the bridges yet?" The bridges on the Danube were perhaps half a mile away, maybe less. You couldn't see them from here, but they were world-famous for their beauty, according to Russell's Baedecker.

The man in the hat shrugged. "Soon." He reached into his leather knapsack and produced a package wrapped in a napkin and a large clasp knife. He unwrapped the package and, without dropping anything, he deftly cut a few slices of salami, onion, and black bread with one hand, and handed them around. That was a trick I knew. My father and my uncle Zoli could do that, and often did—it must have been something that recruits were taught in the old Hungarian army.

David Hurn/Magnum Photos

It was good, and brought back childhood memories. We all had a swig of *barack* to wash it down. "We liberated lots of that too," the one-legged man said, with a grin. The girl took a swig along with the rest of us. She was some girl. From the direction of the river we could hear the fight intensifying—rising above the noise of the artillery was the sharper, earsplitting crack of tank guns firing, and short bursts of machine-gun fire.

"The guys holding the bridges are getting fucked in the ass right now," our friend with the wooden leg said. The girl did not smile, but neither did she seem offended.

"Will they be able to hold them?" Attila asked.

"Not for long, poor sods. Anyway, the city is surrounded. The Russkis will come from every direction, not just across the bridges."

"Still, it would mean something, to hold the bridges. Once the Russians have them, they'll attack the Parliament building, and seize Nagy."

"Yes, probably." He put his finger against his head, made a pretend pistol of it, and pulled the trigger. "And then, bang, he's a goner. You know, I thought he was just another communist stooge, with soft hands. Always talking about 'the workers.' What the fuck does he know about work? But now, I have a certain respect for the guy. He has balls." He gestured toward the blond girl. "Wouldn't you say he had balls, honey?" She nodded. "You see? Blondie thinks so too. Nagy discovered them a little too late, maybe, but better late than never." He lit a thick black cigar that looked like a piece of rope; then, turning toward us, he said, "Speaking of balls, where are our friends from America?"

"We're English," I said.

He bowed deeply, doffing his hat in a sweeping gesture. "Forgive me, my lord. And where is the army of our distinguished British friends?"

"On the way to Egypt, it seems," I said.

"Of course. Look after the empire, that's what comes first, for you guys and the fucking French. And then there's our friends the Israelis! One would like to have seen *that*—Jews in tanks! What next? Well, I can't stay here gossiping all day long, so good-bye and good luck. You be specially careful, Blondie, eh?" We tried to ignore this reference to the Russian soldiers' habit of raping women, and waved good-bye as he stumped back to the truck, and it clattered off down the street. Judging from the girl's expression, the danger of rape was something she had already considered.

Attila passed out the tools. The students dug up the cobblestones quickly enough. A few iron railings and a couple of big trucks were added to it. Attila was anxious to add the ornamental streetlights and the poles that hold up the tramlines to it, but these proved difficult to bring down. I suggested Primacord, an explosive in the form of a rope that can be wound around an object and then detonated from a safe distance, having attended a course in which we were instructed in the use of such things by a sergeant major of the Royal Engineers, but the students didn't have any. Eventually a boy on a bicycle was sent off, and brought back some Czech "Semtek" plastic explosives and a package of detonators from a Hungarian army unit. Watching Attila and his intellectuals handle explosives was enough to drive us back to a safe distance, in the doorway of the hotel. However, they eventually managed to bring down a fair number of streetlights and drag them over to the barricade without having killed themselves. By mid-morning, it had acquired a certain bulk, although whether or not it would stop a Russian tank remained to be seen.

The blond girl—her name was Kati, we discovered—and a round-faced boy who didn't seem to be more than twelve had spent the morning collecting empty bottles. What they needed, she said—her English wasn't bad—was gas. Did we have any? Christopher nodded, walked over to the VW, and came back with one of our jerricans.

"I say," Roger said, "we may need that, you know."

Christopher shrugged. "We're not going anywhere for the moment." The truth was that if she hadn't been such a pretty girl, he would probably have been the first to argue against giving up any of our precious gas. The drill was that you dipped a strip of rag in the gas, then balled it up tightly and stuffed it into the neck of the bottle, with one end hanging down. The trick with a Molo-

Jack Esten / Getty Images

Astoria Hotel, before the attack. Christopher Lord may be at the left.

tov cocktail, Kati explained, was that you had to guess the right moment to throw it. If you lit the rag and threw the bottle too slowly, the whole thing went off in your face, burning you badly. We listened with expressions of rapt interest on our faces, and no temptation to pick up a Molotov cocktail at the first sight of a tank.

From the direction of the bridges the sound of fighting was still intense, but the artillery shells were landing closer to us now. The Russians were in no hurry. This time, when they received sniper fire, the infantry simply backed off, waited for the tanks to get there, and blew up the whole building—or, if in doubt, the whole block.

It was hard to ignore the shelling, though most of the shells were landing a good distance from us. Even so, there's an instant, just before a shell lands, when it seems to suck all the air up in its wake. This is an illusion, of course; shelling happens far too quickly for that. But there's a split second of intense quiet, together with the feeling of being in a vacuum, unable to take a breath; then there's the huge sound of the explosion, momentarily deafening you, making the ground tremble beneath your feet, shattering glass in windows all around you, followed by a swiftly rising plume of smoke and debris that fills your eyes, your nose, and your mouth with foul-smelling grit and the acrid stench of high explosive. Then, finally, comes the spine-tingling clickety-clack of bits of red-hot shrapnel, shattered stone, and odd pieces of plumbing landing. But all this takes place in a fraction of a second, really; it's just that the mind slows things down.

And it isn't happening just once; it's repetitive, constant, mind-numbing. Sometimes the shells land in the street, sending up lethal showers of cobblestones and tram tracks, and then geysers of mud, steam, and water from ruptured pipelines. More frequently the shells hit buildings, exploding when they've penetrated the roof and plunged on through one or two floors below it, making the whole building suddenly swell outward like a balloon that's been overinflated, then collapse in on itself, finally spewing itself out into the street in the form of smoking piles of rubble. It was quite a thing to watch, probably the unacknowledged spectator sport of the twentieth century.

Of course it gave one pause to realize that some twenty-year-old Russian gunner was pulling the lanyard a couple of miles away without much concern for where the shell would land—there's nothing "personal" about random shelling—but that's the kind of thought one wanted to avoid.

Attila declared a lunch break at noon. I sat down on the barricade with him, and he pulled the usual Hungarian lunch out of his haversack: black bread, a slab of bacon coated with paprika, an onion, a bottle of *barack*. We shared it companionably as the shells whistled and exploded in buildings all around the city. I recalled reading that shell splinters were the major cause of death on the western front in the war of 1914–1918, and now I could see why—occasionally they came flashing past at supersonic speed, glowing bright red, from small slivers that could take your eye out or slice off your ear to great chunks that would cut you in two. Attila ignored them. "I am reading much English literature for pleasure," he said between mouthfuls, his expression serious.

I nodded encouragingly, smiling like an idiot.

"I am reading Mr. Graham Greene."

I said that I too was a fan of Graham Greene, and a friend.

"So? That is interesting. Would you call him a progressive writer?"

This would have been hard to answer even without the noise of close and distant gunfire. I explained that Graham was certainly progressive in some ways, but not in others. He was in favor of left-wing governments in foreign countries, particularly those we would now call "third world"; but at the same time, in England he was a traditionalist, and part of what we called the "establishment"—a rebellious and bloody-minded member of it, to be sure, but a member all the same. I tried to explain what the establishment was, conscious that I was not making much headway.

Attila considered all this thoughtfully. "So," he said. "This is like our '*nomenklatura*' under the communists, yes? But tell me, where do you put a fascist, like Mr. Evelyn Waugh?"

I was momentarily stumped by Attila's pronunciation of "Waugh," which came out as "Vaough," but once I understood who he was talking about, I explained patiently that Waugh was not a fascist at all—he and Greene were Catholic converts, close friends, and fellow members of the upper middle class, but Waugh was a reactionary. Since both of them had a taste for schoolboy practical jokes, it was possible that Waugh's reactionary views and Greene's leftist views were merely different forms of a put-on, intended to infuriate and mislead people they disliked. To both of them, class, beauty, and intelligence mattered more than politics.

This was too much for Attila to absorb, coming, as he did, from a place where there was a political or ideological explanation for everything.

"You have to understand," I said, "that in England nobody takes politics seriously. In England, politics are a hobby, like collecting model railway trains—or collecting miniature liquor bottles, which Graham Greene does."

This was a novel, and unacceptable, idea to Attila. He had turned against communism, but he had been educated as a communist. "But *Brighton Rock* is a political novel—a proletarian novel," he protested.

"No, no, it's about Catholicism, not about politics. . . . Take Pinky . . . " I stopped in mid-sentence, having noticed that the firing had lessened in front of us in the direction of the Danube, and increased behind us, to the east, where the Russians were presumably making quicker progress. Either way, it seemed possible that the barricade would be attacked from one or both sides very shortly—not a cheerful prospect. I pointed this out to Attila, who waved it away impatiently. "Whichever side they come from, we will fight them, I think," he said.

At just that moment, Ian and Martha appeared, dressed as if for lunch at Claridge's. They were on the way to the American legation, where there was to be some kind of press conference, and asked if I would like to join them. Why not? I thought. There did not seem to be much to be gained from conducting a PEN seminar on the contemporary English novel with Attila until the Russians arrived to kill us.

I made my apologies and got into their car, and we took off through streets that unarmed civilians and armed fighters were busily barricading. At each barricade we were stopped and asked for our papers, as well as for an explanation of why our army was going to Egypt, instead of coming here.

"Chap I know," Ian said, "told me this whole show was all Eden's bright idea. Encourage the Hungarians to rise up and pin down the Red Army here in the streets of Budapest while we go for the canal."

"Surely not?" Eden seemed to me old, sick, tired, and bewildered by the postwar world, a man who had waited too long for

Churchill to retire and had succeeded to power years too late, rather than a Machiavellian manipulator of whole countries. I had met him once at one of my uncle Alex's parties. I don't know what he was doing there; he had a certain distrust of Churchill's "monied" friends, people like Max Beaverbrook, or Brendan Bracken, or Alex, who had kept the Churchill bandwagon rolling through good times and bad; perhaps Eden's young wife, Clarissa, enjoyed a touch of film glamour. In any case, though still handsome, Eden had looked drained, and older than I had expected. He was like a lot of Churchill's political collaborators: such dynamism as he had once possessed came from Churchill, who had a remarkable ability to fill those around him with his own energy and optimism. On his own, no longer heir apparent but prime minister, Eden had looked to me seedy and slightly deflated, but then we did not know that he was already ill.

Ian was an admirer of Eden's—it was his suits, perhaps, which Ian's rather resembled. "Oh, it's quite possible, you know," he said. "The Russians believe it's true, and they're steaming mad about it. The Yanks too. They think Eden tried to snooker them."

"He won't succeed," Martha said.

"No, no, I agree. If we'd seized the canal right away, it might have worked, but it's all moving much too slowly. . . . If Winston were still P.M. the paratroopers would have seized both ends of the canal before the Gyppos could block it. . . . As it is . . . I say, here we are."

We got out and sauntered into the American legation. Most of the Western journalists except the Dalrymples had given up any attempt to look dressy, and the general effect was that of a damp, unruly mob of refugees, except for Ian, Martha, and the man from the *Times*. The French and Italian journalists were dressed as if for winter sports in the Swiss Alps. The press conference was brief—America was doing its best to broker a peaceful resolution to the problem; the president was in touch with everybody; for the moment the borders were reported to be firmly closed; if things got worse American citizens—but *only* American citizens—would be welcome to take shelter in the legation. This was said with a stern look in our direction.

None of this was exactly earthshaking news, and most of the journalists left in search of something more stimulating. The Dalrymples knew one of the American diplomats, so we stayed behind while they chatted with him, and were therefore among the few people who saw the big news story of the day: Cardinal Mindszenty's arrival at the legation.

He came in through the door, accompanied by a monsignor. Mindszenty was a gaunt, pale-faced figure in a black soutane, with scarlet piping, buttons, and sash. His face had a certain aristocratic grandeur to it, with flashing gray eyes and an aristocratic beak of a nose, but he looked angry and defeated. As it turned out, he knew Martha Dalrymple, and he paused on the big, ornamental stairway to the second floor to say hello to her. They spoke briefly in German, and then he blessed her and went on up the stairs, where a group of Americans waited to receive him on the next floor.

"I say," she said, "there's a story! The cardinal went to the Parliament building to put himself and the church under Nagy's protection, and Nagy told him he couldn't do anything for him, that the best thing Mindszenty could do was to take sanctuary at the American legation. He said Nagy told him, 'I would go there myself, Your Eminence, if I could, but it wouldn't do for an old communist like me.'" There was no love lost between Nagy and Mindszenty. The cardinal was hardly out of prison before demanding the return of all church properties confiscated by the state and the resumption of the traditional role of the Catholic church in education—all things to which Nagy would have been unalterably opposed. The notion of Mindszenty's seeking the protection of a man he despised was a sign of just how bad things looked.

We spent some time at the legation while the Dalrymples persuaded their friend to radio the Mindszenty story out to the West for them, since the telephone lines were cut. Nobody could have been more eager to see us go than the Americans—it was one thing to give Cardinal Mindszenty sanctuary, but they were petrified at the possibility of having to protect a whole crowd of journalists and waifs and strays, many of whom were not American citizens.

And who could blame them? Ian asked, as we got back into our car. Bad enough to have Mindszenty as a houseguest for the next few years! By now it was possible to drive the short distance back to the Astoria Hotel only by stopping at every barricade and seeking information about what was going on around the next corner.

Back at the Astoria, the barricade outside the hotel was crushed and smoldering, the street pockmarked with shell craters and torn up by tank treads, the glass in most of the windows shattered by the concussions. There was no sign of Attila or Kati, or their bodies. The barricade cannot have held the Russians up for more than a few minutes while they shelled it, then sprayed it with machine-gun fire. Curiously enough, our car, with its Union Jack still tied to the canvas roof, was right where we had parked it, and undamaged. Down the street a couple of Russian tanks—they operated in pairs, each one cautiously protecting the other—were parked so as to block the street off. Each turret swung slowly back and forth, with the tank gun elevated as high as it would go. At intervals one or the other would stop and fire toward the rooftops of the surrounding buildings. Clearly they were being directed by somebody, perhaps on the roof of the hotel, to suppress sniper fire. Inside the hotel, there was a lot of plaster dust and powder fumes. Each time a tank fired, the chandeliers tinkled and danced. I made a mental note not to stand under one of them.

The concierge was still at his post. "Your friends are in the bar," he said. We found them there, drinking. They had their heavy coats on—the heating system had failed—and looked shaken. They had been out around the city encouraging the students by keeping a stiff upper lip. Since the Hungarians felt we had betrayed them by our Suez adventure, all the British, including the journalists, felt obliged to set a good example by exposing themselves to danger, as my friends had been doing all morning. When they returned they found that the Russians had dealt with the barricade so quickly that there had been no chance to see what had happened to Attila or Kati. Once the smoke cleared, they were gone.

We sat glumly, drinking and smoking, as the light outside faded. When it was dark, the lights and the heat suddenly came back on. The concierge came through the bar to give us the latest news from the radio. János Kádár had broadcast to announce that a new government, the "Hungarian Revolutionary Worker-Peasant" government, had been formed "to protect Hungary's socialist achievements" and its fraternal relationship with the Soviet Union, and that he was premier.

There were rumors that Imre Nagy and his closest followers had taken refuge in the Yugoslav embassy, and that British and French paratroopers had at long last—far too late—been dropped in Egypt to seize both ends of the canal.

From time to time, the hotel rocked from a nearby explosion, while outside the noise of shooting and the explosions went on all night. There was a dull red glow over the rooftops of Budapest, punctuated from time to time by a brilliant white explosion. The rattle of small-arms fire was dying down.

For all practical purposes the revolution was over.

8.

A Much Fought-Over City

For the moment, it was still possible to walk around the city—the Russians were busy mopping up pockets of resistance, their secret police units were waiting outside the city, and their soldiers were not at first inclined to stand around in the street dealing with civilians or asking to see their identity papers while sniper fire continued. Sensibly, they stuck together, infantry and tanks, methodically attacking any sign of armed resistance. Meals were brought up to them by mobile field kitchens.

It was all done according to the book, with no big surprises. The tanks went down the wide boulevards in echelon, so that the tank on the left could protect the one on the right, and vice versa, each turret swinging slowly back and forth. They avoided the narrow side streets as much as they could, though they sometimes used side streets to outflank one of the larger barricades. Most of the time, when they came to a barricade, they got as close as they could to it, stopped, and fired directly into it. If they could demolish it with their tank guns, they took the time to do so, then moved forward carefully in low gear and ran right over it, crushing it flat, while the infantry came up behind the tanks to polish off any survivors with grenades and submachine-gun fire; if it was too big for the tanks to climb over—some of the barricades relied on tramcars that had been derailed and pushed into place by trucks—they shot it up as much as they could, then provided covering machine-gun fire for the infantry to move forward and deal with anybody who was still alive. Where possible, the infantry fought their way into the buildings on either side of the street barricades, so that they could fire down on the fighters manning the barricades from the windows above and catch them in a murderous crossfire.

This too was standard infantry tactics, nothing fancy. First the infantrymen kicked down the front door of a building—or blew it in if it was solid and old-fashioned, as many in Budapest were—and once they were inside they went methodically from apartment to apartment, kicking in the door, tossing a grenade into the room, then giving it a couple of quick bursts of submachine-gun fire before entering. If a window was still intact, they smashed it to give themselves a clear field of fire, but in most places anywhere close to the fighting the glass had already been shattered by the shock waves from cannon fire.

This was the kind of thing the Soviet Army was good at—there were even still quite a few older Russian officers and senior NCOs around who had done it before in Budapest, when they had besieged the city, then fought their way through it from building to building near the end of World War II. Many of the big apartment buildings and public buildings still bore the scars of the fighting in 1945, and older people whose apartments overlooked a barricade had mostly had the sense to move out before the Russian infantry arrived. This perhaps explains why the civilian death count was comparatively low, given the picture of destruction the city presented. One thing people had learned from 1945—and apparently passed on to the younger generation—was that you didn't want to sit in your apartment guarding your most treasured possessions if the Soviet infantry was at all likely to break down the door and throw a grenade in.

Most of the Soviet tanks were still armed with armor-piercing ammunition designed to penetrate a tank's armor before exploding—there had been no time for crews to unload the ammunition in the turret and replace it with high-explosive rounds. That led to a curious difference between the damage inflicted during the first round of fighting and the second. In the first round, high-explosive tank shells brought down the walls of even the most solid buildings, without necessarily inflicting serious damage on the interior, which was often exposed like that of an old-fashioned dollhouse. In the second round, the armor-piercing tank shells penetrated the exterior walls without exploding, kept going through several interior walls, and then eventually exploded in unpredictable places deep inside a building—just where the residents were, unluckily for them, apt to be hiding as far away from the street as possible.

We heard plenty of stories about being on the receiving end of the shelling. A shell would plow through the thick outer wall, leaving a neat hole not more than six inches in diameter, keep going through the thinner and less solid inner walls, and finally blow up when it hit an object hard enough to trigger the fuse, deep inside the building—a steel beam, perhaps, or a massive, old-fashioned piece of nineteenth-century plumbing. There were people who had been in their room when a shell passed through, and remembered hearing only a deafening crack and a whistle, followed by a storm of plaster dust and stone chips, and a dull explosion deeper in the building that shook everything, even things like the big cast-iron bathtubs which appeared to be unshakable. The passage of the shell through the room took the breath out of the lungs in a terrifying rush, as if all the air had been sucked out in a single microsecond; but when they at last managed to draw a breath again, they realized that they were alive, still sitting unharmed in the same chair or in bed, but covered with dust and debris, ears ringing, dazed, and staring in amazement at two neat holes opposite each other in the walls. Of course for the person several rooms away who was lying in bed when the shell finally exploded, it was a different story, one with no happy ending.

Survival in a city that is being fought over is largely a matter of luck. One of the first things you learn is that ducking, weaving, and taking shelter in doorways were of limited value once the shelling and machine-gun fire began in earnest, though they offered a certain amount of protection against sniper fire, of which there was a good deal. When it came to snipers, we did what everybody else did—we crossed our fingers and made a rush for the next doorway. It was important not to hang about, once you were safely in a doorway—that gave you the illusion of temporary safety, but multiplied the chances that the sniper would adjust his aim and catch you in the open when you made your run for the next doorway. The thing to do was to move fast, and unpredictably, trying not to run in a straight line, and to bear in mind that every time you heard a shot it had either missed you or killed somebody else.

As we soon discovered, everybody had his or her own survival story, and despite the language barrier, everybody was eager to share it with any stranger who would listen. People clustered in small groups, mostly in doorways or on street corners that offered a clear view of any trouble in the making, dusting their clothes off with their hands, and taking a discreet inventory of who was alive and who was not in their building, or on their block. Most people had taken shelter in the basement of their building. Many of the basements still bore air-raid instructions from World War II, and these shelters were a pretty safe bet, given the construction of most of Budapest's apartment buildings—which had been built in the middle to late nineteenth century, in the days when Franz Josef was emperor, and were made to last, with huge blocks of granite, perfectly fitted and carved with decorations by stonemasons who knew what they were doing and were in no hurry to finish the job. It would have taken more than a few hits from a 105-millimeter tank gun to bring these buildings down and trap people in the basement.

Most of the people in the street were elderly, with an air of faded gentility: men in worn but respectable dark suits, stiff collars, and jaunty hats; women in overcoats that had doubtless once been fashionable, with a scarf over their head and worn but decent shoes or boots. They looked like middle-class people of any European country—there was nothing proletarian about them at all. In this respect, too, the Hungarian communist party had failed dismally—despite fierce rhetoric on the subject, dire threats, arrests, the confiscation by the state of small businesses, and the attempted elimination of independent professionals, Hungary still did not resemble the Soviet Union. The old women wearing a flowered hat and a threadbare but nicely fitted overcoat, and carrying a handbag or a string shopping bag as they emerged for a breath of fresh air on Kossuth Lajos Utca near the Astoria Hotel, were unmistakably of the *haute bourgeoisie,* however impoverished—and however much the communist party might claim that all forms of social and class distinction had been eliminated.

An elderly gentleman walked over to us, loden-coated, with an upright posture, a spry little white mustache, and a well-brushed hat—a figure who seemed to belong more to late-nineteenth-century Vienna than to a communist country in the middle of the twentieth century. "You are English?" he asked.

I said we were, all too obviously. He spoke English well, with a strong Hungarian accent, but having grown up around my father, his brothers, and their friends, I was used to that. He had visited England many times, before the war, he said. He knew London well. A very nice city. He had always stayed at the Hyde Park Hotel, where the maître d'hôtel was Hungarian. I seemed to remember my uncle Zoltán mentioning the same thing.

"Zis is terrible, is it not?" he asked, waving a gloved hand toward the broad avenue. To the right of us, in the direction of the Danube, the remains of the barricade outside our hotel were still smoldering. Farther on, Soviet tanks were lined up on the pavement, two long rows stretching for block after block. The tank crews had regained a measure of confidence now—most of the tanks had their hatches open, since no one likes to be cooped up with the hatches closed any longer than is absolutely necessary. In many of the tanks, the commander was exposed from the chest up, elbows on the steel rim of his hatch on top of the turret, scanning the rooflines with binoculars, while the gun loader in the open hatchway beside him followed the direction his commander was looking in with the muzzle of the big .50-caliber machine gun mounted on top of the turret, both hands gripping the big wooden handles, and both thumbs on the trigger bar. The Russian tank crews wore strange-looking padded leather helmets with earflaps, rather like an old-fashioned football helmet from the Knute Rockne era. A few of the tank crews were standing in the street beside their tanks, smoking, sheltered from sniper fire by the solid bulk of the machine.

In the other direction there was a mass of demolished barricades, and the blackened hulks of Soviet and Hungarian tanks and self-propelled artillery hit during the first days of fighting, when it had seemed that the Hungarians were winning and that the Russians might leave. Most of the bodies had been cleared away, but there were still a couple of pitiful ones farther down the avenue, outside a grocery shop or bakery, people who had been hit and killed while they were standing in line for bread or bacon. At intervals, from that direction, a firefight broke out from time to time, causing the tank drivers to gun their engines noisily, and the tank turrets to swing around toward the sound, like large animals alarmed by a noise.

Neither the Russian officers standing around a Soviet jeep nor the Hungarian civilians around us seemed much concerned by the sound of fighting a couple of blocks away. It had been the same way with the people queuing up for food—nobody wanted to lose a place in the line by taking shelter; and if somebody was hit, he or she was simply dragged out of the line to the nearest available cover, if alive, or, if dead, left at the edge of the pavement to be picked up later by whoever was gathering bodies. I had come to understand that this wasn't from selfishness, or callous indifference—if your family needed food (and whose did not?), your duty was to stay in line, however frightened you might be. And, eventually, for the civilians as for the fighters themselves, the sheer closeness of the fighting—the fact that it was taking place right next to you, perhaps 100 feet or less from your front door or from the neighborhood bakery, and threatening to engulf you at any moment—led to a kind of numb, stoic courage. Ignoring the noise of battle and the various lethal chunks of stone and metal that were flying through the air became something of a badge of honor, as much for the old women with flowered hats as for the Soviet officers and noncoms standing around their command jeep in the middle of the avenue, calmly chatting and smoking their cigarettes in full view of snipers, ignoring the amount of unpleasant stuff that came flying through the air, or the occasional ground-shaking thud of an artillery shell landing nearby. It was rather like the way New Yorkers carefully pay no attention to a wailing ambulance or fire engine as it hurtles by them, except that what people were ignoring here could kill you.

The elderly Hungarian had the same exquisite manners and dazzlingly white teeth as so many of my father's Hungarian friends. It had always seemed to me that Hungarian men came in two distinct types: the ones with perfect manners who kissed hands and the ones who were rude, gruff, and disheveled and never kissed hands. My father was of the second type, as was my uncle Zoli, but many of their friends were not. Both types were capable of great charm, but behind the charm was always a trace of unconcealed self-interest and a ruthless determination to win even the smallest of contests or bargains, typical of a people whose public life had been dominated for centuries by Turks or Austrians, and who saw themselves as a tiny minority obliged to scheme for survival against the hostile and vastly larger populations of Slavs that surrounded them.

I offered my newfound friend a cigarette. He took it and held it carefully for a moment, as if it were a rare object of great value, then drew from his pocket a short, carved amber cigarette holder; put the cigarette in it; and inclined his head as I lit both our cigarettes. "Ah," he said, exhaling smoke. "A Players. You have no idea how long since I smoked one.... I used to bring them back from London in the old days, when I went there on business trips." He made a gesture with his hands as if holding something round. "In sealed tins, you know?"

I knew. My father also bought sealed tins containing 100 cigarettes. I asked if he had gone to London often.

He shrugged. "Before 1939, certainly, many, many times. I was in the leather business—luxury leather, for purses, handbags, jewelry boxes, bookbinding, and so on. Morocco leather, the finest calfskin, that sort of thing.... Then the war came, and we had to make leather goods for the German army. An army uses a lot of leather, you know; still, it's not the same.... Then, after the communists took over, we were, how you say, 'nationalized'? Anyway, our company was taken away and turned into a state enterprise. On the other hand, I cannot complain. Many of our competitors were Jewish, and what happened to them in 1944 and 1945 was much worse."

"Was the fighting in 1945 worse than this?"

"Allow me, please, to say: much, much worse. Well, the Germans were ordered to hold out here to the last man, you know. The city was declared a 'fortress'—*Festung Budapest.* The Führer forbade any surrender, so you can imagine.... The siege lasted one hundred days. How we longed for the end of the war, and cursed the Allies for being so slow.... In the end, we didn't care that the Russians were going to take the city instead of the Americans and the British, we just wanted it to be over.... Of course who could have predicted that it would turn out like this?"

"Who indeed?"

"Well, it is nothing new.... Budapest is a much fought-over city. Just in this century, there was Béla Kun in 1918, then the Romanians, then Horthy, then 1945, and now this.... As soon as one gets the city tidied up there's another war or revolution. Not like London."

"London had the blitz."

"So. But not with the tanks in the streets, or firing squads, or bodies hanging from trees. Or mass rape. Being bombed is bad, of course, but being invaded by foreigners is much worse, believe me, my dear sir." He took the cigarette out of the holder, and carefully pinched it until it was extinguished, then slipped the stub in his pocket. "I didn't catch your name," he said.

When I told him he raised an eyebrow. I was never quite sure how people would react to my name in Hungary. Some people—perhaps most—simply admired the Kordas because they had become rich and famous abroad, which for many remained the Hungarian dream. Older people remembered the films my uncle Alex had made in Hungary, and his marriage to my aunt Maria, then a great star until talkies brought her career to an end; but I couldn't depend on a friendly reaction. To Horthyites of the old school a Hungarian Jew, however famous, was hardly Hungarian at all; and the fact that in 1919 Alex had accepted an appointment to Béla Kun's "Directory for the Film Arts," which reported to the Council of Commissars, was neither forgiven nor forgotten—for Kun was an ardent believer in the value of the film as communist political propaganda, and lavished on the Hungarian film industry a degree of personal attention and interest which might better have been directed toward arming the proletariat or fighting the Romanians.

Alex had always said that he had had no choice, if he wanted to continue making films, and I suspect that this was true. For the same reason, my father joined the communist-led Trade Union of Artists and Designers in 1919, during the Béla Kun period, and as a result was expelled from the Academy of Fine Arts when Horthy seized power; he left shortly afterward for Vienna. There was a price to be paid in twentieth-century Hungary for making a political mistake, even in the arts—one of Alex's friends, the film director Sándor Pallós (who also happened to be Jewish), was tortured and executed by Horthy's followers for having made a movie of Gorky's *Chelkash,* which, unfortunately for Pallós, Béla Kun admired. It is hardly surprising that so many Hungarian film people fled to Hollywood, where the worst that could happen to you was being fired by Louis B. Mayer or Harry Cohn, not being tortured and shot. On the other hand, convinced communists objected to Alex's famously expensive lifestyle, his friendship with reactionary figures like Churchill, and the many films he had made glorifying the British empire. To them Alex had sold out and had put himself, objectively, on "the wrong side of history"—the "right" side being presumably represented by the idling tanks on Kossuth Lajos Utca.

Whatever the elderly gentleman with the cigarette holder thought, he was at any rate too polite to say it. He shook his head in wonderment. "Of all things," he said. "What a time to come looking for your—" He paused for the word. "Roots?"

"I'm not really looking for them."

"No? But it might be an interesting thing to do. You'd have to start in—where?"

"Pusztatúrpásztó, where they were born. Or Turkeve, where they went to school."

"A good, solid rural background. In Hungary, this is considered an excellent thing. All Hungarian politicians claim to have grown up on a farm, with bare feet and dirt under the fingernails. But your father's family were Jewish, were they not?"

I nodded.

"Then you wouldn't find too many of your relatives there alive today, I'm afraid. The Germans went through that part of the country late in 1944, rounding the Jews up. The SS were very thorough, and the Arrow-Cross people and the Hungarian gendarmerie very brutal." He said it matter-of-factly. Here, the murder of the Jews was just another horror of war and politics, not a special one, or, as in some places in the West, the only one people knew about.

"So I've read."

"You're very well informed about Hungary, for a young Englishman. And your friends and colleagues, do they also know any Hungarian history?"

"No. Most of them don't have a clue about anything that happened here before the students tore down Stalin's statue last week."

"This is the problem with a small country like Hungary. Nobody cares about our history. We know more about your history than you know about ours. But we Hungarians have a thousand years of history, after all—quite as long as yours. More history than we really need, frankly. Most of it is tragic, unfortunately."

He looked up and down the avenue. Not much seemed to be going on. In a nearby telephone booth there was a pile of arms—Freedom Fighters who were abandoning the fight or were wounded often left their weapons in a telephone booth, for somebody else to pick up and use. By now, however, all those with a grain of sense were getting rid of whatever weapons they had. Once the Russian secret police and military police arrived, and with them the members of the AVH who had survived, possession of a weapon would almost certainly lead to a death sentence, or perhaps even summary execution. This was not, after all, a war—it had been a revolution superimposed on a civil war, and in revolutions and civil wars people don't carry a copy of the Geneva Convention about with them in their pocket. There were no rules on either side about surrendering. AVH personnel and informers who had come out into the street with their hands up had been shot at point-blank range by infuriated armed civilians; and the Russians viewed the Freedom Fighters as reactionary franc-tireurs whom they were perfectly entitled to execute—a view that would have been shared by most armies. Even under the Geneva Convention an armed civilian is not entitled to the same treatment as a uniformed soldier. This was something that every student or worker who took up arms had to face, not to speak of Hungarian army personnel, who would be treated as traitors for turning their arms on the Russians and the Hungarian secret police.

Of course the time to start thinking about this sort of thing is not when the other side has just won; but all over Budapest—indeed all over Hungary—people were grimly facing the price they would have to pay for having taken part in the revolution. Soon, there would be arrests, investigations, forced confessions, a careful scanning of the names of those who had signed petitions or made broadcasts and speeches, and no doubt a painstaking search through the thousands of photographs that had appeared in the Western press for faces that could be identified and matched up to a name. There is always a terrible price to be paid for a revolution that has failed, and the Hungarians were about to pay it.

Across the street, my friends were chatting with a small group of students, now unarmed, under the sharp but indifferent gaze of the Soviet tank crews—only two things interested them at present, weapons and cameras, and neither was present. I had long since stopped carrying a camera around, for nothing was more certain to provoke them. Somebody in the Kremlin must have complained to the Soviet defense ministry about the number of photographs of the fighting appearing in the Western press, and the complaint had made its way down to the level of the troops.

My Hungarian friend and I shook hands with central European formality. Since the streets looked clear, he was going to fetch his wife and take her shopping, he said. There would probably not be much to buy, he thought, but it was important to stick to one's routine. I gave him the rest of my packet of Players, which at first he refused to accept, than slipped into his pocket gratefully. "It will remind me of happier days, before the war," he said. "Go careful, English."

"And you too."

"We are not making politics, my wife and I, so I think we have nothing to fear. But things will be very hard here for a long time." He shrugged. "This we are used to in Hungary, however."

I walked over to the telephone booth—it bore a certain resemblance to the ones in London—fished through my pocket for change, and called the number that Graham's friend the MI6 man had made me memorize in the bar at the Ritz Hotel. To my surprise the telephone worked. I could hear it ringing at the other hand, despite a lot of static and buzzing. It took an age before anyone answered, but eventually a crisp English voice said, "British legation."

"I was given this telephone number by a chap in London," I started to explain.

"Given it? It's in the bloody Budapest telephone book."

"I was told that this was a confidential number that I was to memorize, not write down."

"Nonsense. There's nothing confidential about it all. Who is this, anyway? We're frightfully busy here."

I gave our names, and was swiftly interrupted. "Not the Oxford undergraduates, surely? You chaps had better get over here as quickly as you can. There's been a hell of a fuss about the lot of you. The Hungarians have been raising hell, and so has the Foreign Office. What on earth did you think you were doing, wandering around Budapest in the middle of a revolution? Is Judy Cripps with you?"

"No."

"Damn. Nobody knows what's become of her. I do think you might have realized that we have more important things to do than to wet-nurse a bunch of undergraduates over here on a lark."

"We were delivering medicine . . ."

"Oh, very noble of you, I'm sure. Well, don't expect any thanks from the Foreign Office. Or your colleges, either, I shouldn't wonder. I suppose you know the border's sealed?"

"Nobody told us that."

"Well, I'm telling you. Sealed tight as a drum. Get over here, and smartish, please. If there's one thing we don't want it's a show trial of British subjects."

This was very far from the kind of greeting that I had expected, but my comrades didn't seem particularly surprised or concerned. We decided to walk over to the legation—the students felt we would be safer on foot than in the car, since the Russians tended to open fire at any moving vehicle, even the ambulances, which they had, mistakenly, convinced themselves were carrying ammunition and reinforcements to the fighters in the streets.

There was something inexpressibly sad about the level of destruction. It's not something you notice while the fighting is in progress—the noise, the adrenaline rush, and the smoke obscure what is happening, and produce a kind of tunnel vision, in which you look only at what is directly in front of you, or at the place from which danger seems to be coming. (Of course that can get you killed. Staring at the place from which machine-gun bullets appear to be coming tends to immobilize you, like a rabbit caught on the road in a car's headlights at night, and makes you an easier target to hit.) In any case, you don't waste a lot of time studying the effect of artillery fire on nineteenth-century buildings while the shells are still falling—it's only afterward that you realize the extent of the damage, and the sheer, wanton carelessness of war.

That and the smell of war—the odors of burned rubber, of oil and gas, of damp earth and cellars that were never meant to be exposed, of plaster dust and brick powder from the mounds of debris, and of cooking gas and sewage from severed mains and half-demolished houses. The city, as we walked through it, seemed deserted, except for clusters of Soviet troops at strategic intersections, but of course that was an illusion. The revolution was over; the safest place to be was indoors, waiting for "order to be restored," as the phrase goes.

At places where the fighting had been particularly intense—certain of the more important intersections on the broad avenues leading toward the Parliament building, the Kilián barracks (where Pal Maleter, who was then a colonel, had led the fight), and around the Parliament building itself—the destruction was enormous, proof, if any had been needed, of just how fierce the fighting had been. Masonry and rubble had poured out onto the cobblestones from damaged buildings; tanks and self-propelled artillery that had been hit by gunfire were scattered around in profusion; everywhere there was broken glass and twisted tram rails and snarls of cable like the most grotesque modern sculpture, and the omnipresent sharp chemical smell of expended high explosives. Paving and sidewalk pavement were scarred by tank treads and explosions, and in many places the cobblestones had been pulled up to create a barrier, as outside the hotel. There was no telling how much of Budapest had been destroyed, but walking through the main streets of the central part of town was a deeply depressing experience.

It was readily apparent why the Russians had won so quickly. They had, of course, a huge advantage in numbers, and an overwhelming superiority in weapons; and despite later claims to the contrary, the insurgents had no such thing as a unified insurgent strategy or command. Indeed, the different insurgent groups had no way of contacting each other or a rebel "headquarters," had one existed outside the Russians' imagination, except the telephone system, whereas the Russians were linked by a radio in each vehicle, and were following a well-thought-out plan. The Freedom Fighters threw up barricades where they pleased, while the Russians advanced methodically from point to point, first sealing off the city, then taking the bridges over the Danube, and finally moving from major intersection to major intersection along the broad avenues of the city center, eliminating points of resistance and isolating the fighters in the streets from any contact with what remained of Nagy's government. If nothing else, they had learned something from their failure in the first phase of the revolution—and doubtless from wiser heads at home in Moscow who had fought their way into Budapest against the Wehrmacht, the Waffen SS, and Hungarian fascist forces in 1945.

The British legation was cordoned off by Russian tanks, but this, we were told by one of our student guides, was because of a demonstration by Hungarian women, who had gathered outside the legation to sing patriotic songs and the Hungarian national anthem and had been dispersed by Russian tanks. Nobody seemed clear about why the women had picked the British legation, though it seemed to me possible that they might have been intending to express shame and outrage that the British had not intervened in any way—and by attacking Egypt had freed the Soviet Union to use force in Hungary—rather than any kind of solidarity with the West. Whatever they were hoping to communicate, the Russian tank crews had been determined to prevent them from doing it, and one of them had even aimed his tank gun straight into the doorway of the British legation to discourage anybody inside from trying to make contact with the Hungarians outside.

The tanks had by this time been reinforced by NKVD men, posted to prevent anybody but British subjects from entering the legation. Our passports were laboriously examined and our names taken before we were allowed in. Inside the door, the British military attaché glared at us, every bit as hostile as the Soviet security troops outside, and examined our passports suspiciously in the hallway. "Well, I suppose you're British," he said reluctantly, handing them back to us. He looked at us from head to foot as if we were recruits on parade. "God knows you look British enough." He sighed. "You've put us all to a good deal of trouble," he said.

That had certainly not been our intention, I pointed out. Things had been pretty bad out there in the streets, in case he hadn't noticed.

"Things have been pretty bad in here," he replied sharply. "We've got nearly a hundred odds-and-sods camping in the legation. Journalists, photographers, waifs and strays, even a few mothers with infants.... People have been sleeping on the floors, the canteen has run out of food, and every bathroom is in use. We ran out of sherry days ago. The legation isn't a hotel, you know. We can't feed every British subject in Budapest, and certainly not for free—the F.O. would never authorize that."

With varying degrees of resentment and irritation we said that we weren't expecting to be fed—when I had called, I was told we should get over here on the double, and here we were.

"Who told you that?"

He hadn't given me his name, I said.

"Well, he should have. And you should bloody well have asked."

I could see that there was going to be an explosion from either Christopher or Russell. Christopher had a certain amount of regimental pride and swagger—at any moment he was likely to ask what the military attaché's regiment was, and point out that his own was vastly senior and more distinguished—while Russell, though lethargic and even-tempered most of the time, was capable of sudden bouts of violent bad temper. Tactfully, I suggested that we leave the hallway and go inside, and as we did, I saw that the military attaché was at least correct about conditions inside the legation—it looked like a Gypsy encampment. It had, to begin with, the typically ambivalent foreign office decor—the building must once have been a luxurious private mansion, but the furniture was strictly Ministry of Works "G-Plan" stuff, cheap plywood imitations of Scandinavian modern. Glittering nineteenth-century chandeliers sparkled above fake wood tabletops and plastic chairs with chrome legs; a lovely marble stairway, designed for making a grand entrance, curved up to a cheap-looking wooden door marked "Private"; walls that had no doubt once borne expensive paintings were covered with cork bulletin boards.

Beneath the chandeliers an untidy, unhappy, and unruly band of British passport holders was encamped. There were photographers, surrounded by their equipment; journalists clutching their portable typewriters as if a typewriter were a badge of office; harassed mothers trying to cope with screaming infants; and many unidentifiable types with some claim to British citizenship who had surfaced in Budapest and taken refuge from the fighting in the legation. A few were stretched out on the floor sleeping, but most of them were engaged in noisy recrimination with the legation staff—the journalists wanted to communicate with their papers in London, the photographers wanted to get their negatives out of the country as quickly as possible, the mothers wanted fresh milk and diapers, and the staff wanted—more than anything else—to get rid of these unwelcome guests. None of us was tempted to join these British subjects in distress, whose complaints could not be satisfied—there was no milk in Budapest; the Russians had cut the international telephone lines; nobody knew what was happening at the border crossing points to Austria and West Germany.

"You see how it is," the military attaché said, and we did. The squalor of the scene was enough to kindle a certain sympathy in our hearts for him and the long-suffering legation staff. He led us through the crowd of our fellow Britons, ushered us into a tiny office, and handed us over to a harassed diplomat, perhaps the one I had talked to.

"I don't know how or why you got into the country, or what you've been doing," he said brusquely, "but how do you propose getting out?"

"The same way we got in," Roger said.

"No. Out of the question. The Hungarians and the Russians would love to get hold of you four. They'll say you were sent here to stir up the students."

"That's all rot."

"I daresay, but they'll get somebody to say that's what you were doing. Even get one or two of you to admit to it. No, no, don't underrate them, not when it comes to that sort of thing. . . . You'll have to leave in a convoy. We're trying to put the first one together now. Oh, and by the way, take that bloody Union Jack off your car! The Hungarians and the Russians think it's a sign that you have some kind of official status. You can stay here if you like until we've got permission from the Russian *Kommandtura* to send the convoy."

It did not seem to surprise the diplomat that we chose to return to the hotel, which, however badly damaged, would surely be more comfortable than the legation. Perhaps pleased that he was not about to get four more boarders, he agreed to let us know when the convoy was forming, probably early tomorrow morning, and sent us on our way with a warning not to make trouble.

But trouble was the last thing we had in mind. We and the remaining British journalists sat in the bar of the hotel in our overcoats drinking warm beer by candlelight—electric power was intermittent and spotty. There was nothing cheerful to say. János Kádár had announced on the radio that the new government was in control of the situation, although from here and there in Hungary a few heartbreaking farewell transmissions were being recorded from places where the insurgents, now being labeled "fascist counterrevolutionaries and criminals," were still in control. These would cease within the next few hours, as one by one the last voices of the revolution were silenced.

For the first time in many days there was silence in the streets of Budapest as well—no shots, no explosions, no sound of tank treads. Armed resistance had ceased.

Early in the morning a messenger from the British legation came around to tell us where the convoy was being formed, and gave us all little printed stickers to put on our windshields.

Everyone else had the word "journalist" printed in English, Hungarian, and Russian. Ours had the word "students" written out in ink.

At the appointed time we assembled in a broad avenue near the legation, a motley collection of cars and buses—mostly buses, or cars with Austrian plates, which the journalists had rented in Vienna. Roger, in a burst of patriotism and a reluctance, typical of the Royal Navy, to give in to orders from a mere civil servant, had insisted on keeping the Union Jack on the roof of the car. It caused a certain amount of consternation among the journalists, some of whom thought it might protect us from being strafed, while others thought it might inspire the Russians to shell us.

Our friend from the legation read us our marching orders. We were not to leave the convoy, which would stop at appropriate intervals for certain inevitable "human needs"; under no circumstances were we to attempt to take photographs; we were not to talk to civilians or Soviet military personnel; we were to have our passports ready to show at all times. The convoy would be led by a car from the British legation, and followed by a car from the Hungarian Foreign Office—a glance behind us revealed a shady-looking trio of heavies in shiny black leather trench coats and the kind of hats that Jean Gabin used to wear in French gangster films, standing beside a muddy Russian Pobieda sedan. Christopher waved at them, but they ignored him. "Thugs," he said.

Taking our place in line, we drove slowly out of Budapest. If the center of the city was depressing, the western outskirts were worse—lines of Soviet tanks, army trucks, and armored infantry carriers stretched for miles. Moscow had sent an army of 500,000 men to subdue a nation of 10 million people, of whom only a few thousand, many of them university students and teenagers, were armed and fighting. And even so, the insurgents had soundly defeated the first Russian attack and had held the massive second one off for several days of pitched street combat. It was not exactly a military performance calculated to make the men in the Kremlin sleep better at night.

Few civilians were in the streets, and those we saw carefully avoided looking in our direction. The day was gray and misty, with occasional showers of cold rain—autumn in central Europe. There had been little or no fighting in the dreary suburbs of Budapest—this was hardly surprising, since most of the fighters had been university students or factory workers—and except for the occasional Soviet patrol on the road, there was not much in the way of a Soviet presence as one left the city. No doubt, in their usual methodical way, the Russians had seized the airports, the railway stations, the telephone exchanges, and the radio stations. Once these were firmly in their hands, the rest could be left to the secret police and the communist puppet government of János Kádár.

As we proceeded toward the border, we could see, on either side of the road, long lines of people walking in the same direction—men pushing bicycles loaded with their belongings, women pushing strollers and prams, small children, older people carrying suitcases or sacks thrown over their shoulders, an endless procession of misery.

It was impossible not to feel guilty at the sight of them. We, after all, were in a car, and at the border we would be handed over to the Austrians—they were on foot, and would have to find a way across the border as best they could. Without passports or visas or money they would become stateless refugees the moment they left Hungary, and would be unable to return for many years, if ever. They were, in effect, "voting with their feet" against communism, determined to leave the country at any cost or risk. No doubt some of them had fought in the revolution and had a good deal to fear once "order" was restored and the secret police resumed work; but many—by the looks of them, the majority—had not, certainly not the women with small children, or the old people, or the small groups of children walking toward Austria by themselves, so far as one could judge.

Győr, when we drove through it, was a picture of desolation—the fighting here had been intense, and everywhere there were signs of it: bodies, burned-out tanks, shattered buildings with gaping holes in the masonry, the remains of desperate last-minute barricades. We did not stop—indeed at every intersection there was a Russian vehicle, and a group of soldiers with sullen, angry faces waving us on toward the West.

At the border, we did not get out of our cars. A man from the legation gathered up our passports, took them into the Hungarian border control post, where they were stamped, and brought them back. Somebody had already thought to provide a new rubber stamp with the hammer and sickle on it, to replace the one which had been defaced and which had been used to mark our entry into Hungary. The flag with the communist emblem cut out of it had been replaced too, and there were a couple of Russian tanks nearby, the crews smoking as they stared at us. The refugees, presumably, spread out into the fields and woods before they reached the border crossing and made their way to Austria through open country—a risky business, since there were minefields here and there, as well as plenty of barbed wire.

There was confirmation of this on the other side of the border, as we crossed into Austria. Once they had made their way across the border, the refugees were directed to the Austrian border post, where, drawn up along the highway, there were Red Cross ambulances, army field kitchens, buses, and police vehicles. The Austrians had improvised, with international help, a kind of makeshift disaster relief organization on their border with Hungary, though judging by the number of people milling about they were stretched to the breaking point. Clearly, the Russians had decided not to bother with the frontier for the moment—they had plenty to do in the rest of the country, and besides, there was probably something to be said for letting the malcontents go. For a brief period the Iron Curtain would be open here a crack; no doubt in a few days, when the Russians and the Kádár government had time to deal with it, it would be lowered down tightly again. In the meantime, so long as the refugees stayed away from the official border crossings, nobody was shooting them as they fled from Hungary into Austria and West Germany. It was perhaps the only humane decision made by the Kremlin and the Kádár government in the aftermath of the revolution.

As we drove slowly through the crowds of refugees lining the road to Vienna, I saw quite suddenly, and unmistakably, Graham Greene's friend from MI6, with whom I had had a drink only a few days ago in London. He was dressed in the uniform of a major of the Royal Army Medical Corps, with a Red Cross armband on the left sleeve of his tunic.

I waved at him, but he cut me dead.

9.

"Slaves We Shall No Longer Be!"

With hindsight, all things are clear. It is clear enough now, half a century later, that the resentment of the Hungarian people toward their government and toward the Russian occupiers was deliberately kept simmering by Radio Free Europe, as a kind of reflex taunting of the Russian bear by America; that nobody in Washington ever seriously considered what the consequences might be; and that the British contributed to this situation, with a combination of cunning and ineptitude, as a prelude to their own attack on Egypt—a cynical policy that was to lead to failure on the Nile and the Danube, and marked the demise of the United Kingdom as a great power. Still, the revolution was brewing, and would surely have boiled over without any encouragement from the West.

It is equally clear that the Russians never had any intention of reaching a compromise solution with Nagy. The events of 1956 in Hungary were, in Soviet eyes, simply a fascist "counterrevolution"; to admit even the possibility that they represented a popular uprising was heresy. The Soviet Union stalled and played for time and then, when everything was ready, used massive force—500,000 men; 5,000 tanks; and a full complement of KGB troops—to crush the revolution. Every vestige of doubt, revolt, or disloyalty was zealously stamped out. There were hundreds of executions and many thousands of prison sentences as the Kádár regime took control of the country on behalf of its Soviet overlords. Imre Nagy, who had taken refuge in the Yugoslav embassy, was betrayed by his hosts; kidnapped by the KGB; taken to Romania, where he was tortured; and then brought back secretly to Hungary and shot, like Pal Maleter, along with four of the senior members of his government. Their "trials" and their death sentences were not announced until 1958, with the cynical note that "the death sentences had already been carried out."

Kádár, the supreme realist—who had been arrested and tortured during Rajk's trial, had joined Nagy's government in October 1956, and had then defected to the Russians and presided over Nagy's execution—ruled until 1988, gradually introducing "goulash communism," in which a consumer society made up for the lack of political freedom and choice. Hardly anybody, even in the communist world, ever had a more cynical or longer-lasting career than Kádár. He was the Hungarian Machiavelli, a man with an unrivaled gift for survival. If he had any kind of insight, his last years must have been lived in spiritual torment.

More than 100,000 Hungarians crossed the frontier into Austria and Germany in 1956—whether because the Soviets were happy to see them go, or because the Soviets were too busy with other problems to pay much attention to securing the borders, is hard to say.

In the end—though of course there is no "end" in history—Hungary has become a democracy with a prosperous capitalist economy, and a member of the European Union and of NATO.

This outcome would have seemed unimaginable in 1956, and perhaps was not at all what most of the insurgents really intended. Hungary is, at last, decisively part of the West, and the fault line between East and West has moved, once again, farther eastward, perhaps to the border between the new Ukraine and Russia, perhaps even farther. Volkswagens are being produced in new, modern Hungarian factories; Prada, Vuitton, and Ralph

Topham / The Image Works

Khrushchev greets Kádár in Moscow after the suppression of the revolution.

Lauren shops glitter on the main shopping streets of Budapest; everybody seems to have a cellular phone; and the restaurants include some of the best and most expensive in Europe. A happy ending, one might say, if one believed in happy endings.

Still, the end of the Hungarian story was never in Hungary. The Hungarians stood up to the Soviet Union, bravely and alone; and although they lost, inevitably, they created a deep fissure in the monolith of communism that would eventually—within thirty years, not a long time in terms of history—bring the entire edifice toppling down. It was the story of David and Goliath all right, but with a delayed ending. From 1945 to 1956, the Soviet Union seemed all-powerful, and indeed many people thought its future would be to dominate the world—"We will bury you," Nikita Khrushchev told the Americans—but from 1956 to its inglorious collapse, communism and the Soviet Union were haunted by what had happened in the streets of Budapest, which no amount of space flights or nuclear weapons could paper over. Goliath had been defeated once. Admittedly, he recovered and came back for a second round with a vengeance—but he could no longer lay claim to invincibility, to the "inevitable triumph" of socialism, let alone to humanity and decency. Communism, to borrow the immortal words of T. S. Eliot, died "not with a bang but a whimper," and it was the Hungarians who first challenged its ability to survive.

As for the rest of the story, the four of us who went there (together with Judith Cripps) eventually returned home and resumed our life as undergraduates.

Attila, the university professor turned street fighter, survived, and turned up at a lecture I gave in Toronto. Another one of those Hungarians who can assimilate themselves anywhere, he now teaches literature in Canada.

His girlfriend Kati survived too, stayed in Hungary, married, and sends me a card whenever one of my books is published in Hungary.

I saw Graham Greene's friend from MI6 walking down the opposite side of Jermyn Street some years later, carrying a tightly furled umbrella, and dressed in a dark suit with a bowler hat and—improbably—an I Zingari tie. Our eyes met briefly, but we didn't greet each other.

For those who survived, life goes on, and even looking back in time, for those of us who were there, the events of October and November 1956 seem remote now, something out of another age.

But it is always that way with revolutions. Those who live look back on them with wonder, amazed that they could ever do, or experience, such a thing, as Wordsworth looked back on the French Revolution and wrote:

Bliss was it in that dawn to be alive,
But to be young was very heaven.

Perhaps that is the best epitaph of all.

ACKNOWLEDGMENTS

I am deeply indebted to many people for their help, but would like to thank particularly Carol Bowie and Helen Chin for their help with the manuscript and Kevin Kwan for photo research. I am especially grateful to David McCullough, Michael Beschloss, Lord Thomas, Sir Martin Gilbert, Sir Alistair Horne, and Andrew Roberts for their thoughtful reading of the manuscript and their many practical suggestions; and to Katalin Bentley and Ambassador Gabor Horvath for their help in obtaining documents and research material from Hungary. For their unfailing enthusiasm, I would also like to thank Hugh Van Dusen and Marie Estrada of HarperCollins; and for her absolute faith in my ability to write this book—finally, after all these years—my agent, Lynn Nesbit.

NOTES

CHAPTERS 1 AND 2

I have relied for the most part on *A Concise History of Hungary,* by Miklós Molnár, for historical information; and I found most of the information on famous Hungarians in "Nobel Prize winners and famous Hungarians," HipCat@Hungary.org, a veritable mine of names, biographies, and facts about famous Hungarians.

CHAPTER 3

I have relied largely on Miklós Molnár and on A. J. P. Taylor's *The Origins of the Second World War* for this chapter.

CHAPTER 4

1. Winston S. Churchill: *The Second World War,* Vol. 6, *Triumph and Tragedy,* p. 198.

2. Among the many books consulted for this account of Soviet policy after the death of Stalin, one of the most useful is *The Soviet Tragedy,* by Martin Malia.

3. Csaba Békés, Malcolm Byrne, and János M. Rainer: *The 1956 Hungarian Revolution: A History in Documents,* p. 181.

4. Ibid.

CHAPTER 5

For the chronology of the Hungarian Revolution I have relied, beyond my own memory, on Békés, Byrne, Skardon, and Rainer, *The 1956 Hungarian Revolution*; and often more on Reg Gadney's *Cry Hungary! Uprising 1956,* which is very usefully organized on a day-by-day basis. Where there are differences, I have, for the most part, accepted Gadney's version of events.

1. For the account of events to which I was not a witness, from October 23 to October 30, I have relied principally on Reg Gadney's account in *Cry Hungary!*

2. Békés, Byrne, and Rainer, *The 1956 Hungarian Revolution,* p. 268.

3. Ibid.

4. Gadney, *Cry Hungary!* (and elsewhere).

CHAPTER 6

I am indebted to David McCullough and to Andrew Roberts for their suggestions and help with this chapter.

1. For much of the information and chronology of the next three pages I am indebted to Reg Gadney's invaluable *Cry Hungary!*

2. These are the only names in the book I have changed, except for those of Freedom Fighters.

3. Békés, Byrne, and Rainer, *The 1956 Hungarian Revolution,* p. 354.

4. Ibid, 336.

BIBLIOGRAPHY

Békés, Csaba, Malcolm Byrne, and János M. Rainer (eds.): *The 1956 Hungarian Revolution: A History in Documents, Central European University Press, Budapest, 2003.*

Churchill, Winston S.: *The Second World War,* Vols. 1 to 6, Cassell, London, 1948–1954.

Elliott, Geoffrey, and Harold Shukman: *Secret Classrooms,* St. Ermin's, London, 2002.

Gadney, Reg: *Cry Hungary! Uprising 1956,* Atheneum, New York, 1986.

Gyorkei, Jeno, and Miklos Horvath (eds.): *1956—Soviet Military Intervention in Hungary,* Central European University Press, Budapest, 1996.

Heller, Andor: *No More Comrades,* Regnery, Chicago, 1957.

Horthy, Nicholas: *Memoirs,* Lightning Print, La Vergne, Tenn., 2000.

Malia, Martin: *The Soviet Tragedy,* Free Press, New York, 1994.

Márton, Endre: *The Forbidden Sky,* Little, Brown, Boston, 1971.

Molnár, Miklós: *A Concise History of Hungary,* Cambridge University Press, Cambridge, 2001.

Taylor, A. J. P.: *The Origins of the Second World War,* Atheneum, New York, 1983.

PHOTOGRAPHY CREDITS

AFP / Getty Images: 75

Camera Press / Retna: 143

Jack Esten / Getty Images: 8-9, 132-33, 166

Getty Images: 73, 77, 88-9, 160

David Hurn / Magnum Photos: 163

Michael Korda Personal Collection: Frontispiece foreground, 137

Erich Lessing / Magnum Photos: Frontispiece background, 110, 146-47

Mirrorpix: 149

Popperfoto / ClassicStock: 93

John Sadovy / Time Life Pictures / Getty Images: 102

Sipa Press: 106

Time Life Pictures / Getty Images: 101

Topham / The Image Works: 203

INDEX